Animal Population Ecology

Animal population ecology comprises the study of variations, regulation, and interactions of animal populations. This book discusses the fundamental notions and findings of animal populations on which most of the ecological studies are based. In particular, the author selects the logistic law of population growth, the nature of competition, sociality as an antithesis of competition, the mechanism underlying the regulation of populations, predator–prey interaction processes, and interactions among closely related species competing over essential resources. These are the notions that are considered to be well-established facts or principles and are regularly taught at ecology classes or introduced in standard textbooks. However, the author demonstrates that these notions are still inadequately understood, or even misunderstood, creating myths that would misguide ecologists in carrying out their studies. He delves deeply into those notions to reveal their real nature and draws a road map to the future development of ecology.

T. ROYAMA is well known for his field studies of great tit and spruce budworm, and his contribution to theoretical ecology through the innovative application of stochastic processes. His previous book, *Analytical Population Dynamics* (Chapman & Hall, 1992), had a significant impact on population ecology. He was also a recipient of a Gold Medal of Entomological Society of Canada in 1994.

ECOLOGY, BIODIVERSITY AND CONSERVATION

General Editor:
Michael Usher, *University of Stirling*

Editorial Board:
Jane Carruthers, *University of South Africa, Pretoria*
Joachim Claudet, *Centre National de la Recherche Scientifique (CNRS), Paris*
Tasman Crowe, *University College Dublin*
Andy Dobson, *Princeton University, New Jersey*
Valerie Eviner, *University of California, Davis*
Julia Fa, *Manchester Metropolitan University*
Janet Franklin, *University of California, Riverside*
Rob Fuller, *British Trust for Ornithology*
Chris Margules, *James Cook University, North Queensland*
Dave Richardson, *University of Stellenbosch, South Africa*
Peter Thomas, *Keele University*
Des Thompson, *NatureScot*
Lawrence Walker, *University of Nevada, Las Vegas*

The world's biological diversity faces unprecedented threats. The urgent challenge facing the concerned biologist is to understand ecological processes well enough to maintain their functioning in the face of the pressures resulting from human population growth. Those concerned with the conservation of biodiversity and with restoration also need to be acquainted with the political, social, historical, economic and legal frameworks within which ecological and conservation practice must be developed. The new Ecology, Biodiversity, and Conservation series will present balanced, comprehensive, up-to-date, and critical reviews of selected topics within the sciences of ecology and conservation biology, both botanical and zoological, and both 'pure' and 'applied'. It is aimed at advanced final-year undergraduates, graduate students, researchers, and university teachers, as well as ecologists and conservationists in industry, government and the voluntary sectors. The series encompasses a wide range of approaches and scales (spatial, temporal, and taxonomic), including quantitative, theoretical, population, community, ecosystem, landscape, historical, experimental, behavioural and evolutionary studies. The emphasis is on science related to the real world of plants and animals rather than on purely theoretical abstractions and mathematical models. Books in this series will, wherever possible, consider issues from a broad perspective. Some books will challenge existing paradigms and present new ecological concepts, empirical or theoretical models, and testable hypotheses. Other books will explore new approaches and present syntheses on topics of ecological importance.

Ecology and Control of Introduced Plants
Judith H. Myers and Dawn Bazely

Invertebrate Conservation and Agricultural Ecosystems
T. R. New

Risks and Decisions for Conservation and Environmental Management
Mark Burgman

Ecology of Populations
Esa Ranta, Per Lundberg, and Veijo Kaitala

Nonequilibrium Ecology
Klaus Rohde

The Ecology of Phytoplankton
C. S. Reynolds

Systematic Conservation Planning
Chris Margules and Sahotra Sarkar

Large-Scale Landscape Experiments: Lessons from Tumut
David B. Lindenmayer

Assessing the Conservation Value of Freshwaters: An International Perspective
Philip J. Boon and Catherine M. Pringle

Insect Species Conservation
T. R. New

Bird Conservation and Agriculture
Jeremy D. Wilson, Andrew D. Evans, and Philip V. Grice

Cave Biology: Life in Darkness
Aldemaro Romero

Biodiversity in Environmental Assessment: Enhancing Ecosystem Services for Human Well-Being
Roel Slootweg, Asha Rajvanshi, Vinod B. Mathur, and Arend Kolhoff

Mapping Species Distributions: Spatial Inference and Prediction
Janet Franklin

Decline and Recovery of the Island Fox: A Case Study for Population Recovery
Timothy J. Coonan, Catherin A. Schwemm, and David K. Garcelon

Ecosystem Functioning
Kurt Jax

Spatio-Temporal Heterogeneity: Concepts and Analyses
Pierre R. L. Dutilleul

Parasites in Ecological Communities: From Interactions to Ecosystems
Melanie J. Hatcher and Alison M. Dunn

Zoo Conservation Biology
John E. Fa, Stephan M. Funk, and Donnamarie O'Connell

Marine Protected Areas: A Multidisciplinary Approach
Joachim Claudet

Biodiversity in Dead Wood
Jogeir N. Stokland, Juha Siitonen, and Bengt Gunnar Jonsson

Landslide Ecology
Lawrence R. Walker and Aaron B. Shiels

Nature's Wealth: The Economics of Ecosystem Services and Poverty
Pieter J.H. van Beukering, Elissaios Papyrakis, Jetske Bouma, and Roy Brouwer

Birds and Climate Change: Impacts and Conservation Responses
James W. Pearce-Higgins and Rhys E. Green

Marine Ecosystems: Human Impacts on Biodiversity, Functioning and Services
Tasman P. Crowe and Christopher L. J. Frid

Wood Ant Ecology and Conservation
Jenni A. Stockan and Elva J. H. Robinson

Detecting and Responding to Alien Plant Incursions
John R. Wilson, F. Dane Panetta and Cory Lindgren

Conserving Africa's Mega-Diversity in the Anthropocene: The Hluhluwe-iMfolozi Park Story
Joris P. G. M. Cromsigt, Sally Archibald and Norman Owen-Smith

National Park Science: A Century of Research in South Africa
Jane Carruthers

Plant Conservation Science and Practice: The Role of Botanic Gardens
Stephen Blackmore and Sara Oldfield

Habitat Suitability and Distribution Models: With Applications in R
Antoine Guisan, Wilfried Thuiller and Niklaus E. Zimmermann

Ecology and Conservation of Forest Birds
Grzegorz Mikusiński, Jean-Michel Roberge and Robert J. Fuller

Species Conservation: Lessons from Islands
Jamieson A. Copsey, Simon A. Black, Jim J. Groombridge and Carl G. Jones

Soil Fauna Assemblages: Global to Local Scales
Uffe N. Nielsen

Curious About Nature
Tim Burt and Des Thompson

Comparative Plant Succession Among Terrestrial Biomes of the World
Karel Prach and Lawrence R. Walker

Ecological-Economic Modelling for Biodiversity Conservation
Martin Drechsler

Freshwater Biodiversity: Status, Threats and Conservation
Dudgeon

Joint Species Distribution Modelling: With Applications in R
Ovaskainen and Abrego

Animal Population Ecology

An Analytical Approach

T. ROYAMA
Canadian Forest Service (retired)

CAMBRIDGE
UNIVERSITY PRESS

CAMBRIDGE
UNIVERSITY PRESS

University Printing House, Cambridge CB2 8BS, United Kingdom

One Liberty Plaza, 20th Floor, New York, NY 10006, USA

477 Williamstown Road, Port Melbourne, VIC 3207, Australia

314–321, 3rd Floor, Plot 3, Splendor Forum, Jasola District Centre, New Delhi – 110025, India

79 Anson Road, #06-04/06, Singapore 079906

Cambridge University Press is part of the University of Cambridge.

It furthers the University's mission by disseminating knowledge in the pursuit of education, learning, and research at the highest international levels of excellence.

www.cambridge.org
Information on this title: www.cambridge.org/9781108844420
DOI: 10.1017/9781108951135

© T. Royama 2021

This publication is in copyright. Subject to statutory exception and to the provisions of relevant collective licensing agreements, no reproduction of any part may take place without the written permission of Cambridge University Press.

First published 2021

Printed in the United Kingdom by TJ Books Limited, Padstow Cornwall

A catalogue record for this publication is available from the British Library.

Library of Congress Cataloging-in-Publication Data
Names: Royama, Tomo, 1930– author.
Title: Animal population ecology : an analytical approach / Tomo Royama, Canadian Forest Service (retired).
Description: Cambridge, UK ; New York, NY : Cambridge University Press, 2021. | Series: Ecology, biodiversity and conservation | Includes bibliographical references and index.
Identifiers: LCCN 2020042223 (print) | LCCN 2020042224 (ebook) | ISBN 9781108844420 (hardback) | ISBN 9781108948166 (paperback) | ISBN 9781108951135 (epub)
Subjects: LCSH: Animal populations. | Animal ecology.
Classification: LCC QL752 .R69 2021 (print) | LCC QL752 (ebook) | DDC 591.7/88–dc23
LC record available at https://lccn.loc.gov/2020042223
LC ebook record available at https://lccn.loc.gov/2020042224

ISBN 978-1-108-84442-0 Hardback
ISBN 978-1-108-94816-6 Paperback

Cambridge University Press has no responsibility for the persistence or accuracy of URLs for external or third-party internet websites referred to in this publication and does not guarantee that any content on such websites is, or will remain, accurate or appropriate.

To the memories of my mentors
David E. Lack   Shunro Utida
George, C. Varley   Patrick A. P. Moran
Nikolaas Tinbergen   R. Franklin Morris
Charles S. Elton   Maurice E. Solomon

Contents

Prologue *page* 1

1 Hunting Strategies of Predators as Revealed in Field Studies of Great Tits 7
  1.1 Preamble 7
  1.2 Search Image Theory by Lukas Tinbergen 8
  1.3 Alternative Theory: Hunting by Profitability 11
  1.4 Profitability Curve 21
  1.5 Allotment of Hunting Time among Different Sites 23
  1.6 Hunting by Profitability as Principle 25

2 The Paradox of Crypsis: Is it Effective against Visual Predation? 26

3 Logistic Law of Population Growth: What Is It Really? 31
  3.1 Preamble 31
  3.2 The Classical Logistic Equation 32
  3.3 Fundamental Nature of Population Processes 34
  3.4 Ecological Significance of the Differential Equation: $dx/dt = xf(x)$ 34
  3.5 Discrete-Time Processes 39
  3.6 Classical Logistic Model as Particular Case of Model (3.10) 46
  3.7 Reinterpretations of Parameters ρ and K 48
  3.8 Structual Problem of the Common-Version Logistic Model 50
  3.9 Final Remarks of Chapter 51
  Appendices
  Appendix 3A: How to Solve a Differential Equation in the Models (3.1) 52

Appendix 3B: The Derivative $d(e^u)/du = e^u$ 55
Appendix 3C: The Derivative $d(\ln v)/dv \equiv 1/v, v > 0$ 56
Appendix 3D: The Anomaly in the Common Version Logistic Equation (3.3c) 56
Appendix 3E: Mathematical Attributes of the Verhulst Equation (3.3a) 56

4 Reproduction Curves and Their Utilities 58
4.1 Original Ideas 58
4.2 Drawing a Reproduction Curve 61
4.3 Generating the Population Series 61
4.4 Mathematical Roles that the Model Parameters Play 63
4.5 Problems with Population Size as a Non-negative Quantity 66
4.6 Logarithmic Transformation of a Reproduction Curve 68
4.7 An Application to Actual Data 71
4.8 Variation in Dynamical Pattern of the Model Process (4.4b) 75
4.9 Examples of Variations in Dynamical Pattern 78
4.10 Difference between Discrete-Time and Continuous-Time Processes 82
4.11 Ecological Feasibility of Variations in Discrete-Time Processes 83
4.12 Endogenous and Exogenous Processes 85
4.13 Application of an Endogenous–Exogenous Process Model to Wildlife Management 89
4.14 The Origins of the Myths of the Logistic Law 90
4.15 Final Remarks of Chapter 91

Appendices
Appendix 4A: The Derivative of a Function Is a Measure of the Slope of the Curve Generated by the Function 92
Appendix 4B: The Derivatives of a Few Standard Functions 95
Appendix 4C: L'Hôpital's Rule 96
Appendix 4D: Prototype Curve and Its Translation 96

5 Generalization of the Logistic Model 98
5.1 Preamble 98
5.2 Negative Binomial Distribution 100

	5.3 Ecological Application of the Negative Binomial Distribution	105
	5.4 A General Model of Intraspecific Competition	108
	5.5 Model (3.10) as a Particular Case of Model (5.12)	112
	5.6 Interpretation of the Hassell Model: $r_t = x_{t+1}/x_t = r_m/(1 + ax_t)^b$	113
	5.7 One More Model to Examine	113

Appendices

Appendix 5A: Why $0! = 1$? — 115
Appendix 5B: Why the Name 'Negative Binomial'? — 115
Appendix 5C: How to Calculate the Mean and Variance of the Random Number m in (5.4) — 116
Appendix 5D: Why Do the Terms jk^{j-1} $Qr(j)$ in (5.6) Vanish in the Limit $(j \to \infty)$? — 118
Appendix 5E: Convergence of the Sum $\{\Sigma[(h + j - 1)!/h!(j - 1)!](kq)^{j-1}\}$ to $(1 - kq)^{-(h+1)}$ — 119

6 Scramble and Contest Competition: What Is the Difference? — 120

   6.1 Preamble — 120
   6.2 Drawing Reproduction Curves Based on Model (5.12) — 121
   6.3 Broader Interpretation of Parameter h — 126
   6.4 In the Weirdland of a Negative Hit — 128
   6.5 Nature of Competition — 130
   6.6 What Determines Parameter h in Actual Processes? — 132
   6.7 Scramble and Contest as Elements of Competition — 136
   6.8 Concluding Remarks of Chapter — 140

Appendices

Appendix 6A: The Logarithm of a Negative Real Number is a Complex Number — 140
Appendix 6B: How to Estimate Parameters (R_m, h, c/h) to Fit Model (6.1a) to the Observed Reproduction Curve in Figure 6.3a — 143

7 Regulation of Populations: Its Myths and Real Nature — 145

   7.1 A Brief History — 145
   7.2 Biological Population Processes As Stochastic Processes — 146

	7.3	Defining Population Persistence	150
	7.4	Investigations into Mechanisms for Persistent State	151
	7.5	Density-Dependent Processes under Exogenous Influences	153
	7.6	Density-Independent Processes	154
	7.7	Algebra of Stipulation (7.3) for Population Persistence	158
	7.8	Random Walk As Unregulated Processes	161
	7.9	Density-Dependent Regulation	162
	7.10	Precise Nature of Density-Dependent Regulation	167
	7.11	Density-Independent Regulation	169
	7.12	Logical Problem in Climatic-Control Theories	170
	7.13	Myth of Density-Dependent Regulation	174
	7.14	Concluding Remarks of Chapter	177

Appendices

Appendix 7A: Rules of Operations on the Expectations Used in the Present Chapter — 177
Appendix 7B: Derivation of Relationship (7.5) — 178
Appendix 7C: Calculation of an Autocovariance Function (ACVF) — 180

8 Predator–Prey Interaction Processes — 181

8.1	Preamble	181
8.2	Formulation of Endogenous Models of the Interaction Processes	181
8.3	Simulation of the Dynamics of Predator–Prey Interactions	186
8.4	Variation in Dynamical Patterns	190
8.5	Effects of Random Exogenous Influences	197
8.6	Reproduction Surfaces of a Predator–Prey Process	206
8.7	Revealing Conditional Reproduction Curve in Observed Series	212
8.8	Problems Inherent to Earlier Models	217
8.9	Interactions between Predator Complex and Prey Complex	219

Appendices

Appendix 8A: Ecological Mechanism Underlying the Equation $q(x_t) = [1 - \exp(-bx_t)]$ in (8.2) — 221

Appendix 8B: How to Find the Equilibrium Levels of the X and Y Series in the Simultaneous Equations (8.4) — 222

Appendix 8C: How to Generate Correlated Series of Random Numbers — 223

9 Interspecific Competition Processes — 225
9.1 Preamble — 225
9.2 Formulation of Competition Model — 225
9.3 Simulations — 227
9.4 Criteria for Coexistence and Elimination — 228
9.5 Reconsideration of the 'Competitive Exclusion Principle' — 245
9.6 Alternative Ways of Viewing Nature — 248
9.7 Struggle for Existence vs Optimization of Profitability — 249

Appendices

Appendix 9A: How to Calculate x^{**} and y^{**} — 250

Appendix 9B: Infeasibility of Category (v) in Table 9.1, Section 9.4 — 250

Appendix 9C: How to Incorporate the Effect of Random Exogenous Influences in the Model — 251

10 Observations, Analyses, and Interpretations: A Personal View through the Spruce Budworm Studies — 252
10.1 An Outline of the Spruce Budworm Studies — 253
10.2 Earlier View of Outbreaks — 254
10.3 Thoughts on the Basic Processes of Ecological Studies — 267
10.4 Concluding Remarks in the Quest for Certitude — 270

References — 271
Index — 273

Prologue

When a university professor retired or died, it often happens that his/her former students get together to compile a festschrift to commemorate their teacher. But, I was not a professor, just a plain research scientist of the Canadian Forestry Service. I had no students who would compile my past works. But that is a good thing after all because, in retrospect, I find that many of my early ideas were neither adequately conceived nor well presented in writing. However, over the later years of my research career, the ideas became consolidated to higher levels of clarity, coherence, and integrity, such that these would now be readily put to a practical use in the study of animal populations. Thus, when I retired from my active research career, I decided to assemble those ideas to share them with younger-generation population ecologists.

The present book comprises ten selected topics as Chapters 1 to 10. Each of these focuses on the subject that is, despite its fundamental importance, still inadequately understood or even misunderstood in the current ecology. I selected only these subjects with which I am familiar through my own personal experiences in the field and theoretical studies. So, these topics cover only specific and restricted areas within the whole realm of ecology. Nonetheless, my perspective presented in this book will hopefully, and in principle, serve as methodological bases for much wider areas of population ecology to be explored by ecologists in general.

Chapter 1 investigates the strategies of predators for hunting their prey, based on my study of the great tit, *Parus major* (Royama, 1966, 1970) with which I started my research career. I found that the birds were able to assess how much food (insects) they could collect (per unit effort of hunting) at a certain site, or how 'profitable' the site would be. Then, I found that the birds allotted time for hunting at a given site according to the profit they could gain there in comparison with the profitability of other sites. This chapter describes how I made observations, interpreted the results, and deduced the idea of hunting by profitability. The idea serves as a basis in Chapters 8 and 9 where I investigate multi-species interaction processes, in particular with respect to the evolutionary process of niche selection by animal species.

Chapter 2 is an extension of Chapter 1 to discuss the evolution of crypsis (or mimesis) among insects as a defensive device against visual predation. Contrary to the common belief that crypsis is effective and therefore has evolved, I found cryptic and well-camouflaged insects comprise a majority of food that the great tits fed to their chicks in the nest: there was little sign that the crypsis, as assumed by these prey species, was effective against bird predation. Here, I attempt to resolve this apparent paradox.

Chapter 3 examines the nature of the logistic law of populations in terms of a theoretical model I develop from first principles, comparing it with the classical model first proposed by P.-F. Verhulst (1838) and later (independently) by R. Pearl and L. J. Reed (1920). The comparison reveals commonly held misconceptions about the logistic law, such as the notion of the 'carrying capacity of the environment', popularly designated by the parameter K. I contend that K is not the environmental capacity. As has already been recognized among mathematicians, it is the stable equilibrium level of the logistic process through competition among the members of the population. The model developed here provides a theory for understanding the real nature of the law and serves as a basis for exploring the principles underlying single-species population processes discussed in a few of the subsequent chapters.

Chapter 4 explores the utility of the 'reproduction curve', a graphical method used by W. E. Ricker (1954) in his analysis of fish populations. I extend the method for general use in animal population ecology. As an example, with this method, I re-analyse the US census data since 1790, which Peal and Reed used for demonstrating the descriptive capability of their model of population growth. My re-analysis provides insight into major factors that governed the US population changes during the past two centuries, demonstrating the potential capability of the reproduction curve to reveal the ecological mechanisms underlying observed population changes. Also, the graphical method exposes some logical problems in the conventional interpretation of the iconic sigmoidal logistic growth curve. In particular, I demonstrate that the sigmoidality has no ecological significance, and suggest an alternative interpretation.

Chapters 5 and 6 examine the nature of intraspecific competition. In Chapter 5, I develop a theoretical model of competition as a generalization of the model developed in Chapter 3. The generalized model enables us to investigate the nature of competition among individuals of a population, the theme of Chapter 6. In this chapter, I look into the nature of scramble and contest competition to show that these are elements of a continuum in that we cannot draw a sharp line between them to distinguish one from the other. Then, I further include sociality as another element of the continuum to show that no sharp line can be drawn between competition and sociality. Thus, we recognize an animal population process as a continuum, made up of competition and sociality as major elements. This recognition serves as a basis for understanding the nature of interspecific competition discussed in Chapter 9.

Chapter 7 critically examines the conventional notion of population regulation as a density-dependent process. I show that the notion has been inadequately defined and caused unproductive controversies in the history of ecology. A fundamental problem is the misconception that density-dependent regulation ensures the persistence of a population. To eliminate the problem I define the notion of 'regulation' and 'persistence' from the stochastic process point of view, and examine the roles played by density-dependent and density-independent processes in regulating a population. In particular, I show that density-independent regulation is theoretically possible but fragile and, hence, would not be realizable in a fluctuating natural environment. On the other hand, I show that density-dependent regulation, too, is defenceless against the adverse effect of environmental changes: that is, the regulation is necessary but *not sufficient* for ensuring that the population persists. A major point of my contention here is to expose, and to rectify, some logical inadequacies and myths in the conventional notion of population regulation.

Chapter 8 develops a theoretical model of predator–prey interactions as a generalization of the single-species process model developed in Chapter 3. The theoretical model generates a variety of dynamical (time-dependent) processes, and I interpret their ecological and evolutionary significance in terms of the variation in the parameter values that characterize the interaction processes. Mainly, I investigate the interaction between one predator species and one prey species to find the principles of the processes. Then, I generalize this one-to-one system to a multi-species system, so as to apply the principles to the study of natural populations in which a complex of predator species interacts with a complex of prey species.

Chapter 9 delves into the interaction between two competing species with particular reference to the controversial notion of the 'competitive exclusion principle' or 'Gause's law'. I first develop a model, similar to the predator–prey process model, but adapted to two-species competition processes. My theoretical investigation here suggests that there are four categories of interactions in accordance with the variations in values of the model parameters, each of which is given an ecological meaning. I show that, in one category, the two species coexist in a stable manner, whereas the elimination of one species is inevitable in the remaining categories. Based on these theoretical studies, I look into multiple-species interaction processes in general and into the ecology (evolution) of niche selection by species in natural environments.

The final Chapter 10 promotes thoughts on the methodology of carrying out an ecological investigation through observations, analyses of data, and interpretation of the results. These are the thoughts that I have conceived while studying outbreak processes of the spruce budworm, *Choristoneura fumiferana*, in eastern Canada. In particular, I critically examine the early ideas (hypotheses) of the processes in which an outbreak is initiated and is spread over wide areas: namely, the formation of epicentres and the dichotomy of epidemic–endemic states of budworm populations. My re-analyses of the original data reveal that these ideas are neither conceivable theoretically nor substantiated observationally. This finding leads me to delve into methodological fundamentals in ecological studies of populations in general: the amalgamation of empirical and theoretical ecology and how to achieve it.

These chapters are arranged in the order that the concepts and methods developed in a given chapter are used (or generalized) in the subsequent chapters. So, I recommend that readers do not skip a chapter, or at least have a quick glance through it to get an outline, at first reading. This is important because, after all, these chapters are concerned with critical examinations of the issues that the readers have most likely been taught at the undergraduate level of ecology. Reading a given chapter in isolation without being aware of the problems that have been dealt with in the preceding chapters, the reader might still be trapped in a pitfall of misconceptions.

I write this book in an informal manner, assuming that I am leading a small workshop with students of ecology at graduate levels, often intervened by questions, answers, and discussions. For example, when I feel additional information might help some readers to readily follow the argument, I insert a [Note: ...] in an appropriate place, even amid the paragraph concerned. That is to say, a [Note: ...] is an answer to a question that is assumed likely to be asked during my talk. Likewise, when I feel a little break might help amid a very involved argument,

I may even insert a [Digression: ...] to relax the readers and encourage them on through the argument.

All chapters but the first two are mathematically inclined. I often build models to be used for the analyses of ecological processes which may require an elementary knowledge of calculus (e.g. how to calculate derivatives or to solve simple differential equations) or certain basic concepts of statistics (e.g. expectation of a random variate or moments of a probability distribution). I do not assume that every reader would readily follow the mathematical arguments involved. So, as these concepts and methods appear in the passages concerned, I explain them in an easy-to-follow manner, either on the spot or in appendices, so that readers do not lose track of the arguments. The only prerequisites for the readers are to have knowledge of a simple algebraic operation like $(a + b)^2 = a^2 + 2ab + b^2$ and a rule like $\log(ab) = \log(a) + \log(b)$ or $a^m a^n = a^{m+n}$.

Any mathematical formula (equation, inequality, etc.) that is going to be subsequently referred to is designated by the chapter number and the order of appearance, e.g. (3.10) is the 10th formula in Chapter 3: if not referred to subsequently, the formula is undesignated. If it is referred to again (e.g. in a different chapter) but I feel that going back to it might interrupt the reader in following the flow of logical argument at that moment of reading, I might *repeat* it with the supplemental acronym 'rpt', e.g. (3.10 rpt) or in another appropriate way.

Similarly, each figure is coded numerically by the chapter number, followed by the ordinal number of the chapter, e.g. Figure 4.3 means the third figure of Chapter 4. The caption for each figure is made as simple as possible such that it serves as a title, so to speak. The full explanations are amalgamated into the argument concerned in the text: in fact, the 'explanations' are part of the argument.

Many chapters have Appendices to help those readers with a minimum background in mathematics to follow the arguments concerned in the main text without losing track. These appendices are coded by the chapter number followed by an uppercase letter, e.g. Appendix 3A means the first appendix in Chapter 3. If mathematical formulae are included, each is coded, e.g. (3A.1) means formula 1 in Appendix 3A.

The bottom line is to bring those readers with limited mathematical skills up to the level at which they can comprehend the ecological meaning of each mathematical model without compromising the level of rigour. After all, I maintain that one cannot do population (or quantitative) ecology without elementary-level mathematics on which the majority of my arguments are based. I realized the necessity of mathematics very early in my research career. But, as I was not a mathematician by trade, I taught myself on the basis of my high school mathematics; I am still learning. The way I write this book has stemmed from my own experience of reading advanced, hard-to-follow textbooks.

My ultimate objective here is to provide the readers, graduate students in particular, with the necessary bases for reasoning and inferences so that they can develop their own methodological skills to tackle the problems that they face in their own research. I suggest that the readers never uncritically believe and accept what I write. Rather, I want them to think with me and digest the ideas. I would feel my mission accomplished if this book successfully motivates young ecologists to venture into the vast and still unexplored realms of population ecology.

Acknowledgements

My words of gratitude are due to the following people. Two old friends of mine, Peter W. Price, Professor Emeritus, University of Northern Arizona, and Stuart L. Pimm, Doris Duke Professor of Conservation, Duke University, acted as my consultants and advisers in preparing the manuscript. Michael B. Usher, Honorary Professor, University of Stirling, and the general editor of the *Ecology, Biodiversity and Conservation* Series, advised me to improve the manuscript in various aspects.

Sara Brunton, a UK-based freelance copyeditor, looked after my English, and my sons, Kenn and Sami Royama, helped me in computer-based technicalities of writing. In particular, Kenn managed, updated and maintained my computer system to suit my needs, while Sami digitally formatted all figures to conform to the typesetter's specification.

Dominic Lewis (Senior Commissioning Editor for Life Sciences) of Cambridge University Press evaluated and accepted my proposed book, while Aleksandra Serocka (Senior Editorial Assistant for Life Sciences) looked after miscellaneous issues, such as handling the copyrighted materials and composing the book covers. Samuel Fearnley (Content Manager at the Press) and Divya Arjunan (Senior Project Manager of SPi Global) coordinated the final stage of the publication process.

I would like to express my sincere thanks to them all for their cordial support and encouragement to make the publication of this (most certainly) final work of mine realized.

T. R.
November 2020, on my ninetieth birthday

1 · *Hunting Strategies of Predators as Revealed in Field Studies of Great Tits*

1.1 Preamble

Towards the end of my undergraduate years at the University of Tokyo, I decided to go to the graduate school in forest zoology where I could study birds in the field. Although called 'forest zoology', in practice it was an entomology laboratory. However, it was the only place where I could possibly study birds in a natural environment. At that time in Japan, field ornithology was not an academic subject in any university.

I proposed to study the breeding behaviour and ecology of the great tit, *Parus major minor* (now classified as the independent species *P. minor*). At the admission interview, one professor (then the head of the University Research Forests) became interested in my proposal and suggested that I use the Fuji Research Forest by the scenic Lake Yamanaka at the foot of Mount Fuji, about 80 km south-west of Tokyo. He also provided me with financial support and with all sorts of facilities at the research forest during my three field seasons from 1958 to 1960.

However, I could not find anybody to guide my study, as there were only a few academic ornithologists in Japan then, and they were mostly taxonomists studying dead specimens. In contrast, there were quite a few works on the tits in Europe, especially in the Netherlands and Britain, which I used as trailblazers. As my study progressed, I recognized certain aspects that had not been well understood even in the European studies. The topic I talk about here is one of the few aspects that I think I understood in depth: the strategy of the great tits hunting their prey (insects). Here I describe how I made observations, interpreted the results, and conceived a general principle underlying the strategy with which predators hunt their prey. But, before getting into the main theme, I review a major theory from the European studies.

1.2 Search Image Theory by Lukas Tinbergen

In the final year of my graduate study in Tokyo, I happened to see an article by Lukas (Luuk) Tinbergen (1960) of the University of Groningen (henceforth just Luuk). In this article, Luuk proposed a theory based on his intensive study of the great tit (*P. m. major*, the subspecies of *Parus major* on continental Europe) feeding their nestlings with various insects in the Dutch pinewoods. In particular, he studied the frequency of a given insect brought to the nest to feed the young in relation to its abundance in the woods.

As a typical example, he selected the case for pine beauty (*Panolis flammea*, Lepidoptera: Noctuidae) caterpillars. What he found was summarized in his figure 14, which contained two curves. One was the frequency (%) occurrence of the *Panolis* caterpillars in the tits' diet and the other the relative (%) abundance of the caterpillars in the woods, both being plotted against their density in the woods. He postulated the following: (i) if a parent tit searched the woods at random, it should have encountered the caterpillars in proportion to their relative abundance in the woods; (ii) then, the % frequency of the caterpillars in the diet should be expected to be proportional to their (%) abundance in the woods. In other words, the two curves in the graph should have matched in shape. Luuk found no match but the following tendency: when the (%) abundance of the insect in the woods was low, its frequency in the nestling diet was below that expected; when the abundance was at a moderately high level, the frequency was substantially above that expected; but at a much higher level in abundance, the frequency did not increase as expected but levelled off below that expected. Luuk interpreted this tendency in the following way.

When a given insect species was sufficiently abundant, the tits encountered it so frequently that they would have been conditioned to its image when searching for prey, i.e. the formation of a search image. This enabled the birds to see the insect more readily among the leaves and twigs that would have otherwise obscured the presence of the well-camouflaged insect. Conversely, when the insect was scarce, the birds did not encounter it frequently enough to form a search image and did not recognize the insect as readily. But, then, Luuk had a dilemma: the search-image hypothesis would not explain why the caterpillars were not brought to the nest proportionately more frequently when they became very abundant. He got around the dilemma by postulating that the birds must have avoided the diet becoming too monotonous with a very abundant prey species. I was not comfortable with Luuk's inferences.

First of all, I was not happy with his postulation of 'random search' as a criterion with which he compared the observed frequency of the prey species in the tits' diet. As is well known, forest insects occur on their own distinct places or niches, such as trees, shrubs, or ground vegetation. Even within each type of place, they tend to occur on different host plants, or even in different parts of the same plant (micro niches). Second, I knew through my own observation that a parent tit as a rule brought home only one prey item on each hunting trip. Under these circumstances, the random search criterion necessarily implies that the tit would visit one of these niches at random in each hunting trip, regardless of the places it had visited on the previous trips. The crucial point here is: would the tits actually do that? It is crucial because, if they did not (and I knew they didn't), Luuk's 'expected' frequencies in the diet would no longer be an appropriate basis on which the formation of search images was deduced. Besides, I was not comfortable with the duality in Luuk's explanation: the formation of search images and the avoidance of monotonous diet.

By the time this paper of Luuk's came out (in fact published posthumously), I had already finished my three-year field study at the research forest in Yamanaka. My routines in the field included mapping the foraging sites of the birds and recording the insects fed to the nestlings; I did the recording at the nests by direct observation as well as with an automatic camera I designed and built. So, I knew that my tits did not search the woods haphazardly for prey. Instead, each tit tended to make a series of trips to a particular site (a tree, a clump of shrubs, or a patch of ground vegetation) and brought home a particular kind of prey in succession. You might say: 'That strongly suggests that the bird formed a search image of the insect after a successful catch'. Well, that is just using the term for conveniently referring to a complex process of learning and memorizing the whereabouts of the things of interest. The formation of a search image that Luuk envisaged was altogether different from the above process of learning. As he described it explicitly, it was a specific (albeit hypothetical) process of recognizing the insect by the bird through its eye among noisy surroundings. It would serve like a template, as it were, of the image of the object formed in the bird's brain by repeated encounters. An image received through the eye would then be recognized by the brain if it matched the template.

In contrast, to understand the bird returning to the same spot repeatedly to look for a particular kind of prey, as I observed, would not have required the assumption of a specific mechanism of visually detecting (recognizing) the prey at the hunting site. All we need to assume is simply that the bird learned where the prey could be readily found. I do not mean that the formation of search images as Luuk envisaged is improbable. All I need to say is: the postulation of the specific mechanism is unnecessary to understand the cause of the trend that Luuk observed, i.e. the pattern of deviation from the 'expected' in the random encounter hypothesis. So, what is an alternative view? Before I delve into it, I would like to note one thing.

I often wondered why Luuk did not consider any alternative to the random-encounter hypothesis before he decided on it. He had firsthand knowledge of tits' searching behaviour in the woods, yet he did not make use of his knowledge. Why? I do not believe that he was satisfied with his own interpretation: it is hard to believe that a researcher of his calibre would have been. He was the youngest of the remarkable Tinbergen brothers: the eldest, Jan, was an economist, a pioneer in econometrics, and second brother Nikolaas (Niko) was, as well known among biologists, a founder of comparative ethology. Both were Nobel laureates in their own fields. Luuk was a professor in zoology, whose earlier works had been highly regarded by his colleagues, but died young at the age of 39. He was very close to his brother Niko, and was greatly influenced by him. Niko was an extremely rigorous thinker: in fact, in later years, I was a regular in his evening student discussion group in Oxford. He never allowed us students to leave even a trace of ambiguity in reasoning; he never allowed us to make an assumption without a logically valid basis. Luuk must also have been influenced by Jan in rigorous statistical inferences. So, it is rather surprising that Luuk did not put more thought into his last work. I think I know why: he took his own life, I am sad to say, and he must have been going through a tough time dealing with the emotional ordeal. So, he could not have put as much thought into the work as he would have otherwise. He left drafts of his work which were published posthumously in two parts: Professor Herman Klomp (a close friend of Luuk's and involved in the publication of the work by Luuk) told me the story when I visited him at the Agricultural University of Wageningen in the early 1960s. I wish I could have met Luuk; I would have had a great time chatting with him.

In the following, I talk about an alternative theory that I conceived during my graduate work in two parts: one at the Fuji Research Forest in Yamanaka, Japan (henceforth the Yamanaka study) and the other at Wytham Woods, near Oxford, UK (the Wytham study).

1.3 Alternative Theory: Hunting by Profitability

1.3.1 The Yamanaka Study

A major objective of my field study in Yamanaka in the summers of 1958–1960 was to find what the tits fed their young with. I recorded 3300 prey items in a total of 210 hours of direct observations at a few nests, and 12,000 photographs on 53 nestling-days at several other nests. But I had a serious problem: I could not identify most of the insects by name, as they were mostly in their larval stages. There were also quite a few adult moths but, before being taken to the nests, their wings had been removed by the parent tits. However, I could distinguish most of them by their characteristic appearance and colour. So, I recorded them by the codes I invented for my own use. I needed an illustrated identification guide of caterpillars, but none was available then. Detailed descriptions of morphological keys for identification in most taxonomic books did not help, as I could not examine each specimen closely in my hand. So, I did not know for sure where these caterpillars occurred in the woods. Another problem was that, by the time Luuk Tinbergen's paper came to my attention in 1961, I had already finished my field work and did not have an opportunity to test his theory. I was only sure that what I had observed in Yamanaka did not agree with Luuk's interpretation.

1.3.2 Study in Wytham Woods

In the late summer of 1962, I obtained a British Council scholarship for further study at the Edward Grey Institute of Field Ornithology (EGI), Oxford, to join its long-term project on the genus *Parus* under the directorship of Dr David E. Lack. There, as David suggested, I concentrated on the hunting behaviour of the great tit (*Parus major newtoni*, the British subspecies of *Parus major*) with a supplementary observation of the blue tit (*Parus caeruleus*, now moved to the genus *Cyanistes*).

The EGI tits studies were being conducted mostly in Wytham Woods, a biological reserve of the University. Also, Professor George C. Varley of the Hope Department of Entomology was conducting his population studies of the oak-feeding insects with his right-hand man, George R. Gradwell. They were interested in my proposed study and allowed me to set up my recorder camera at a corner of their study plot. Also, they provided me with all sorts of information on the insects in Wytham, including their famous work on the winter moth, *Operophtera brumata*, which was in fact one of the most important food sources for the tits. George Gradwell loaned me the Victorian book *The larvae of the British Lepidoptera and their food plants* by Owen S. Wilson with beautifully hand-drawn illustrations by Eleonora Wilson. After a series of on-site tutorials George Gradwell gave me in the woods, I could identify most of the caterpillars. He also provided me with his original population data of the oak-feeding insects. So, I was all set. In Wytham, I took 29,000 photographs on 97 nestling-days over three seasons (1963–1965) at seven nests. Most aspects of the tits' hunting behaviour that I had already observed in my Yamanaka study were confirmed in greater detail in the Wytham study.

1.3.2.1 Key Features in Tits' Hunting Behaviour

I found several key features of the tits' behaviour that provided bases of my alternative theory. First, all individual great tits (both in Yamanaka and Wytham) brought home almost invariably only one prey item on each hunting trip: the closely related blue and coal tits often carried several items. However, except for the first few days after all chicks hatched, each prey item brought to the nest was large; the majority of the caterpillars looked fully mature.

Second, as already mentioned, an individual tit tended to make a number of trips consecutively to the same hunting site (e.g. a particular tree) and brought home the same prey species (a hunting bout, say); but after a while switched to a different prey species in another bout; and so on to repeat the cycle. This tendency is clearly seen in the series of food brought home by a parent tit plotted on a chart over the course of a day (cf. figure 9, p. 647, in Royama, 1970). You would see a run of a particular prey species for a certain length of time, usually not much more than half an hour, only sporadically interrupted by odd individuals of different species, followed by a run of another kind of prey for another time period.

Third, as already pointed out, different caterpillars occurred on distinct plant species. For example, one occurred mainly on oak but another on hawthorn, etc. However, even if occurring on the same tree, the caterpillars therein tended to be resting during the daytime in distinct micro-niches, e.g. leaves, branches, or the trunk of the tree. This suggests that changes in the runs of prey species from one bout to another indicate changes in micro-niches for hunting. Furthermore, each run of a given species often began abruptly, a tendency that happened more often than a run starting with sporadic appearances of a particular species mixed with others. This implies that, at the beginning of each hunting bout, the tits had already decided exactly where to hunt for what prey.

These features convinced me that the tits were very efficient hunters. They must have been able to assess how profitable a given prey species, or a given hunting site, could be. They must have been able to assess it (call it 'profitability') by the amount (e.g. most practically, biomass) of food collected in a given hunting bout. Then, the tits must have frequently been sampling many potential hunting sites to evaluate which site was more profitable than other sites and accordingly to allot their time among those sites to gain most profit. This implies that sampling different sites must be a regularly and frequently conducted routine because of the ever-changing levels of profitability among the potential sites.

I now describe in detail what I saw and examine the factors and conditions that most likely determined the profitability of a hunting site. [Note: I invented the term 'profitability'. As English is not my native tongue, I was not sure if it was appropriate. So, I consulted some colleagues at the Zoology Department. I was encouraged to stick with it: even the *Journal of Animal Ecology* accepted the term without fuss.]

1.3.2.2 Chronological Changes in Profitability and Successions of Prey Composition in Tits' Diet

To describe what I observed in Wytham (where I could identify most of the prey), I conveniently divide the nestling period into three major subperiods by the timing of the occurrence of given prey species in the diet. [Note: For the first few days after hatching, the nestlings were fed with small pieces of food, e.g. spiders or a fragment of a large piece of whatever it was. The fragments were difficult to identify on the film. So, I did not operate the camera until a few days later when the chicks were fed mostly with mature larvae. You have to see it for yourself to believe how big a piece a young chick can swallow. It can swallow an item as big as its head. Incredible!]

Early period (the last week in May). For the first 3 days of operating the camera at one nest in 1963 (cf. appendix table 2, p. 660, in Royama, 1970), the mature larvae of the two similar-sized geometrids, *Operophtera brumata* (winter moth) and *Oporinia dilutata* (November moth; syn.: *Epirrita dilutata*), constituted almost 75% (in number) of all prey items brought home by the parent tits. Although comparatively small in size (20–25 mm long), *O. brumata* and *E. dilutata* combined were by far the most abundant prey species in the woods in this early period.

Of the other items, about 14% consisted of miscellaneous prey, including: tortricoid larvae, mainly *Tortrix vilidana* (green totrix), some spiders, and adult dipterans, including tipulids (crane flies), bibionids (March flies), empidids (dagger flies), and syrphids (hoverflies). The individual sizes of these miscellaneous prey were mostly no larger than that of a *brumata/dilutata* larva. The remaining 9%, however, were the larvae of the two similar-sized geometrids, *Colotois pennaria* (feathered thorn) and *Phigaria pilosaria* (pale brindled beauty); these are combined in one category, the *pennaria/pilosaria* larvae. Each of them was twice the body length, and at least 4 or 5 times in volume, as a *brumata/dilutata* larva. Few other caterpillars as large as or larger than a *pennaria/pilosaria* larva were around in the woods during this period of the season.

Now, the number of *brumata/dilutata* larvae brought home per day steadily declined towards the beginning of June: from 441 on 23 May down to 323 on 27 May; sharply down to 170 on 28 May; further down to 89 on 31 May; only 17 in total during the first week in June; and that was it. I considered two possible causes of the decline: one was certain and the other very probable. First, these larvae began to pupate. Like many other caterpillars, they descended to the ground to pupate just under the duff, meaning they were no longer available to the tits. However, George Gradwell's trap catches showed that their descent for pupation did not begin until 29 May; the number descending peaked 3 days later; and by 10 June, few larvae remained on the trees. In other words, until 28 May, these larvae must have been as available to the tits as they were on the first 3 days. So, the substantial decline in the number brought home from 23 to 28 May could not be attributable to pupation.

The other and probable cause of the decline was the larvae of the other geometrids, *C. pennaria* and *P. pilosaria*, rapidly becoming fully mature. Although much less abundant than *brumata/dilutata* larvae in number, they were individually at least 4 times larger in volume. So one hunting trip by the tit that yielded a single *pennaria/pilosaria* larva would have been worth as much as 4 or 5 trips each with one *brumata/dilutata* larva; remember that, somehow, an individual great tit (be it Japanese or British) brought home only one prey item per trip regardless of its size. As the *pennaria/pilosaria* larvae were rapidly maturing, they became more profitable, and the tits steadily lost interest in the smaller prey, even though they were still as abundant as before. Thus, while the number of the large geometrids brought home increased from 49 on 23 May to

163 on 28 May, the number of small prey (*O. brumata* etc.) declined from more than 500 to less than 200.

Intermediate period. Following the first week of observation, in addition to the *pennaria/pilosaria* larvae, the parent tits began to regularly bring home the mature larvae of several noctuids, including most frequently *Allophyes oxyacanthae* (green-brindled crescent) and less frequently *Orthosia populeti* (lead-coloured drab), *Amphipyra pyramidea* (copper underwing), *Brachionycha sphinx* (sprawler), and *Griposia aprilina* (Merveille-du-jour; syn.: *Dichonia aprilina*). The successions of the prey species in the tits' diet are most certainly attributable to the fact that the larvae of different kinds were becoming as profitable (as they mature) as the preceding *pennaria/pilosaria* larvae. I will now elaborate a little more on this issue.

The first major prey of the season, *brumata/dilutata* larvae, is reported to feed on many varieties of plants. There were at least a dozen eligible trees and shrubs in Wytham, but the oak (*Quercus robur*) was by far the most preferred. The larger geometrid larvae, *pennaria/pilosaria*, are reported to feed on a variety of plants, as do *brumata/dilutata* larvae. But in Wytham, according to George Gradwell, they were fairly rare on oak; they occurred more frequently on other trees or shrubs, such as ash, hazel, or hawthorn. Thus, in the tits' diet, the transition from the *brumata/dilutata* to the *pennaria/pilosaria* larvae means that the tits were actually switching their main hunting sites from oak to non-oak plants.

This switching of main hunting sites by the tits was manifested more clearly in the appearance of the noctuid larvae in the nestling diet after the *brumata/dilutata* larvae were rarely brought home: *Allophyes oxyacanthae* fed mostly on hawthorn (*Crataegus monogyna*) and blackthorn (*Prunus spinosa*); *Orthosia populeti* fed exclusively on aspen (*Populus tremula*); and *Griposia aprilina* fed exclusively on oak. [Note: the *A. pyramidea* and *B. sphinx* larvae reportedly feed on a variety of trees, including oak, hawthorn, and aspen, but their frequency occurrences in the diet closely coincided with the rise and fall of the *O. populeti* as well as the *A. oxyacanthae* larvae. This indicates that *A. pyramidea* and *B. sphinx* were likely collected mainly on aspen and hawthorn.] The simultaneous appearance of certain key prey species in the daily menu suggests that the tits were hunting on a number of different tree or shrub species, switching frequently from one site to another during the day.

Furthermore, during the daytime, some larvae stayed on the leaves (e.g. all these green caterpillars: *Op. brumata*, *Orth. populeti*, *A. pyramidea*, and *B. sphinx*), while others hid elsewhere. For example, *A. oxyacanthae* larvae, which perfectly resemble a piece of bark of the hawthorn, most often rested on the trunk or thick branches, whereas the *pennaria/pilosaria* larvae stayed on small branches with a posture to mimic a twig. *G. aprilina* larvae, although feeding exclusively on oak leaves, were dark-coloured and hid in crevices in the tree trunk during the daytime. In other words, the tits not only made successive trips to a given tree species for a while before switching to a different one, but they also knew what to search for in a specific micro-niche (leaves, branches, or trunk) on each tree.

Late period. After the first week of June, the *pennaria/pilosaria* larvae and aforementioned noctuids in turn began to decline –they were still regularly brought to the nest, but in a reduced frequency. The beginning of their decline coincided more or less with the time when the tits started bringing home a large number of tortricoid pupae, mostly *Tortrix viridana* (green tortrix), together with the larvae of a dozen species, including several *Orthosia* species – *O. incerta* (clouded drab), *O. gothica* (Hebrew character), *O. cruda* (small quaker), but mostly *O. stabilis* (common quaker; syn.: *O. cerasi*) – as well as the geometrids *Erannis leucophaearia* (spring usher; now in the genus *Agriopi*) and *Eupithecia irriguata* (marbled pug), and even the hairy lymantrid caterpillars *Lymantria monacha* (black arches), although in single-digit numbers.

The significance of the above transition in species composition in the tits' diet is that all of these late-occurring prey species fed either exclusively or preferentially on oak; moreover, most of them, as they were green-coloured, stayed on leaves during the daytime. In the meantime, *Griposia aprilina*, even though it was an oak specialist, dramatically declined in the diet: as already mentioned, this caterpillar, being dark-coloured, hid itself in crevices in the oak tree trunk during the daytime. Clearly, the transition of the species composition in the nestling diet was due to a shift in the main hunting sites from largely non-oak (during the preceding period) to oak, especially its foliage. This interpretation can also be extended to explain the following fact.

Incidental catches. As already mentioned, earlier in the season when the tits were collecting a large number of *brumata/dilutata* larvae, they also collected some dipteran adults quite regularly, if only in small numbers. However, when the tits switched their attention from the *brumata/dilutata* larvae on the oak foliage to the non-oak-feeding larvae, the dipteran adults were brought to nest only sporadically. However, during the late period when the tits switched their hunting sites back to the oak foliage, some of the dipterans, mostly the tipulids, reappeared in the diet.

This indicates that the tits must have collected those dipterans as they came across them while searching in oak foliage for the caterpillars as the main source of food for the nestlings. In other words, the dipterans were likely incidental catches because they were obviously a trivial source of food compared with the fat and juicy noctuid larvae. The parent tits most probably ate those small items as they came across them while searching for the main foods to bring home.

Although the foregoing events were fairly easy to comprehend, there was one thing which puzzled me for quite a while.

1.3.2.3 A Puzzle to Solve

The miscellaneous tortricoid pupae, collected by the tits in exceedingly large numbers in the late period, were mostly *Tortrix viridana*. They were extremely abundant at an outbreak level in the years I was working in Wytham; its host tree, almost exclusively oak, was heavily defoliated. As a tortricid (leaf roller), the larva folds back the tip of a lobe of the oak leaf over itself, like a Mexican taco wrap. When disturbed, it swiftly wriggles out of the wrap (forwards or backwards) and drops on a silk, a typical predator-avoidance action of many tortricid larvae. So, they are not particularly easy prey for the great tit, which is not as adept at catching these larvae as its cousin the blue tit.

Now, the *viridana* larvae on the trees had been fully mature at the very beginning of the observation (when *brumata/dilutata* larvae were still being brought home in large numbers) and were more abundant (in my study years) than *brumata/dilutata* larvae with a big margin. Besides, *viridana* larvae shared the oak leaves with *brumata/dilutata* larvae. Nonetheless, during that period of time, the tits brought home *viridana* larvae in only small numbers compared with *brumata/dilutata* larvae. Initially, I thought that it was simply because a *viridana* larva was not an easy prey because of its swift reaction to predators, and that the tits took only those in a prepupal stage when they must have been easier to catch.

Eventually, I realized that the problems were not that simple. The *viridana* larvae began to pupate in the last week of May, at more or less the same time as the *brumata/dilutata* larvae. However, unlike the *brumata/dilutata*, the *viridana* larvae did not descend to the ground but stayed on the leaves (within the 'taco' wraps) to pupate. Therefore, they were still available in large numbers as easy prey.

By the end of May, the majority had pupated. Nonetheless, the tits paid them little attention, but collected them only sporadically and in single-digit numbers per day; they appeared to be no more than incidental catches. All of a sudden, however, the (year 1963) pair of tits began to collect the pupae in large numbers: 35 on 6 June, which jumped to 272 on 8 June, and then a whopping 571 a few days later. It took me a while before I began to comprehend. What I realized was the following. There are two distinct issues to recognize: the profitability of the pupae as an individual species and the profitability of the site (oak foliage) of which the *viridana* pupae were a constituent. Let me begin with the first issue.

Each tortricoid pupa was so small that 7 or 8 of them could be worth just about a single large, mature noctuid larva (e.g. *Allophyes oxyacanthae*). Translating this into 571 pupae, and allowing for my crude estimate, they were worth, on average, only 70–80 large larvae. On the other hand, the tits routinely brought home well over 100 of these large larvae a day. To compensate for the small size of the pupae, the tits had to make many more trips between the nest and the hunting site. Was it really worthwhile allotting any amount of time to collecting the super-abundant but individually small prey? [Remember again that, somehow, the great tits brought home only one item at a time regardless of size.] To investigate this issue, I needed to assess what the difference in profitability could actually have been between the pupae and the regularly occurring larger larvae. But how could I do it? After a while, I found a way: crude, but good enough a trick for the purpose.

I looked at a daily chart of the prey items brought to the nest by one female tit; I picked the female simply because the chart I happened to use was that of the female. Further, I arbitrarily picked a long, uninterrupted run of *A. oxyacanthae* larvae and that of the *viridana* pupae, each with a very short average interval between consecutive trips. I picked those runs for the following reason: during such a run, the tit must have minimized the time spent in activities other than those necessary for collecting an individual prey, i.e. travelling from the nest to the hunting site, finding a prey (very quickly), picking it up, bringing it to the nest, feeding it to a nestling, going back to the hunting site, repeating the cycle, and little else. Then, I divided the number of the prey individuals by the length (in minutes) of each run to estimate the minimum average rate at which a single prey individual could be collected and brought home by an individual tit per unit time.

In the particular runs I examined, 16 *oxyacanthae* larvae were brought home without interruption over 28 minutes (or 5.7 larva/10 min) and 25 *viridana* pupae over 13 minutes (or 19.2 pupae/10 min). Now, a full-grown *oxyacanthae* larva would be 7–8 times larger in volume than a *viridana* pupa, so that 19 pupae would be worth only about 2.5 or so larvae, although more nutritional values could have been packed in a pupa. In other words, hunting *oxyacanthae* larvae for a given length of time (given their abundance on that day) could yield a profit at least twice as much as hunting *viridana* pupae. As crude as it may be, the above calculation suggests that hunting the *oxyacanthae* larvae must have been substantially more profitable than hunting the *viridana* pupae.

With the above result, I further investigated how the tits actually allotted their time in hunting between the two types of prey. I looked into a day (8 June) on which the tits (male and female combined) collected 272 *viridana* pupae. On the same day, they brought home 202 *oxyacanthae* larvae. Applying the rate of hunting per unit time (for a single parent bird), I figured that it could have taken 68 minutes for the pair of tits to collect the 272 pupae, as opposed to 177 minutes to collect the 202 *oxyacanthae* larvae.

Setting aside how to interpret the above estimates in absolute value, the point of interest here is the relative values: about 2.6 times more minutes were allotted to the more profitable larvae than to the less profitable pupae on that day. But, why did the tits bother allotting any of their time to the less profitable *viridana* pupae? Why didn't they concentrate solely on the more profitable *oxyacanthae* larvae? You might say: to avoid a monotonous menu, as Luuk Tinbergen suggested. I doubt it, as I have a more plausible explanation.

1.3.2.4 Allotment of Hunting Time Among Prey Species vs. Among Hunting Sites

An obvious and crucial point to consider is the profitability of a site as a whole, e.g. oak foliage where the *viridana* pupae were just a constituent, albeit a major one. This suggests that the tits must have allotted their time primarily among the productive sites, rather than among the individual prey species: the hawthorn trunk–branch site vs. oak foliage, rather than *oxyacanthae* larvae vs. *viridana* pupae. I had this idea when I recognized the following.

On the hawthorn trunks were mainly *oxyacanthae* larvae and few else; similarly, among the twigs there were *pennaria/pilosaria* larvae and few else. In contrast, by the end of the first week of June, as many as a dozen large species had become available among the oak foliage simultaneously.

In particular, in the oak leaves I sampled, I quite often found, beside numerous *viridana* pupae, large noctuid caterpillars. This prompted me to look closely into the daily charts of prey brought home by the tits (cf. figure 3a and 9 in Royama, 1970). Indeed, I found that a single run of *viridana* pupae was often mixed with large noctuid larvae of oak feeding species. It was particularly noticeable that these large caterpillars tended to occur individually in the run of the pupae but seldom occurred in a series. At first, I thought that these caterpillars were merely incidental finds in the oak foliage where the tits were looking primarily for the pupae. However, this does not make good sense because, as already argued, *viridana* pupae on their own could not have been as profitable a source as *oxyacanthae* larvae. The following is a more plausible interpretation.

Recall again that a parent tit brought home only one food item at a time. Thus, at each trip to the oak foliage, the first thing that the tit encountered would probably have been a *viridana* pupa, but every now and then it could have been a large caterpillar. Because of the sheer number of pupae on the foliage, the odds are that the tits found and collected them in succession, whereas encountering a big caterpillar was more sporadic. Nonetheless, the tit could count on big, juicy caterpillars that occur quite regularly in the foliage. In other words, the tits must have assessed the oak foliage to be as profitable a hunting site as the hawthorn trunk–twigs site where *oxyacanthae/pennaria/pilosaria* larvae were collected. In short, it is likely that the tit became attached to a profitable site and took whatever it came across first, rather than looking for a particular prey species with its image in mind. The crucial question is: how do the tits assess the profitability of a given site?

1.3.2.5 How Do Tits Assess the Profitability of a Hunting Site?
In theory, the profitability can be measured by the amount (biomass) of food the tit can collect there to bring home per unit time. But, pragmatically, how could the tit do it without a stopwatch or weighing scale? The answer is by watching the chicks' state of hunger or satiation.

In the Yamanaka study, I spent 210 hours (mainly in the first year before I built the automatic camera) directly watching what was going on inside the nest boxes. I saw that each time a parent tit came home with food, every chick stretched its neck upwards as high as possible and with the gape wide open; it did this no matter how hungry it was. The parent placed the food item in one of the gapes, apparently at random. If the chick was hungry, it immediately grabbed the food and began to swallow it. The parent watched it until the chick finished swallowing the prey and then went out to collect more. On the other hand, if not hungry, the chick did not swallow the food as quickly; often, it even kept the gape wide open with the piece of food already placed in it. Then, the parent pulled it out of the gape and put it in another gape, and so on, until a hungry chick was found. This had already been noted by the EGI alumna Monica Betts, whose pioneering work on the tits' feeding habit guided me in field observations.

If the chicks were quick to react, the parents kept collecting more food. But when all the chicks became reluctant to swallow the food quickly, the parents relaxed their hunting activities. Because the parents tended to stick to a particular site for a while, they must have been able to assess whether the site was profitable by the intensity of the chicks' demand: hunting at a good site would satisfy the chicks more quickly than hunting at a lesser site. However, the parents must also spend time regularly sampling other potential sites to assess ever-changing levels of profitability in the woods. They would probably do this at a slack time when feeding themselves.

1.4 Profitability Curve

Nobody, I hope, would disagree that the profitability of a prey species or a hunting site must be positively correlated with the abundance of the prey therein. But what sort of relationship is likely expected? An idealization (or oversimplification) would help us to make an inquiry into the underlying principle. So, let us assume that the tit hunts at a given site with no structure (a single niche, e.g. leaves or trunk of a tree) within which prey individuals of just one kind are distributed at random.

Now, we plot the abundance (density) of the prey at the site on the horizontal axis and, on the vertical axis, we plot the average number of prey individuals (mean total biomass) that the tit could collect and bring home in a unit period of time, i.e. the measure of profitability. Let us call the resultant curve 'the profitability curve', as illustrated in Figure 1.1. In particular, the curve starts at the origin (zero density, zero profitability) and increases as prey density increases, but with an ever-decreasing rate of increase. More particularly, as prey density increases, the curve increases monotonically and smoothly to ever more closely (i.e. asymptotically) approach a horizontal line which is the maximum profitability that the tit can gain at the site. The reason for this shape is as follows.

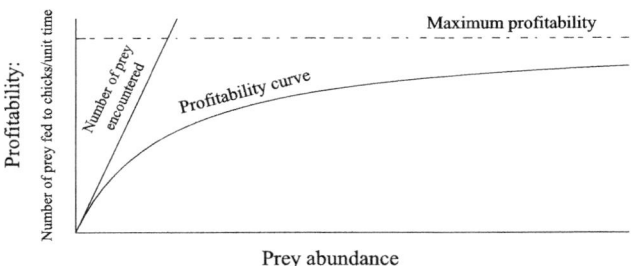

Figure 1.1 Idealized example of profitability curve.

Each hunting trip consumes a certain amount of time required for two major categories of activity: searching for prey within the site and dealing with the prey found. The latter category would typically involve: capturing (chasing if necessary) prey; preparing the catch (killing and removing appendages, if any, e.g. wings); flying back to the nest with it and feeding it to a chick; tending the chicks for a while (e.g. removal of faeces); and flying out to the site to begin searching for another prey.

[Digression: Let me talk about an (extreme) example of how a tit prepared a prey it caught before bringing it home. For some time, I was wondering about a couple of mysterious things in the woods. One was the shrivelled bodies of large hairy caterpillars, like *Lymantria monacha* (black arches), hanging on branches. I just thought they were the carcasses of those that had died of a viral disease. Meanwhile, every now and then, I saw a tit bringing home a blob of dark, shapeless material, looking soft, fresh and juicy; whatever it was, the chicks seemed to love it. Then, one day I saw a tit pick up a fully mature, very hairy *L. monacha* caterpillar in the oak foliage, carry it over to a branch, grab its body with one foot, and repeatedly peck hard at the joint between the head and the first thoracic segment. As the tit decapitated it and pulled the head off the body, out came the whole length of innards. The tit rolled them up neatly into a blob around its beak (just like those I saw fed to the chicks in the nest) and flew away, leaving the empty hairy caterpillar skin hanging on the branch. It made my day.]

Now, just think of a tit searching for prey along a random path, forgetting about capturing the prey for now. That is, just think of the number of prey individuals that the tit would encounter per unit time, given the prey density, each time it searched the site. Because we are assuming a random occurrence of prey individuals at a site with no structure, the tit would be expected to encounter the prey in proportion to prey density: here, I am plagiarizing Luuk Tinbergen's random search. The number encountered per unit time at the site is represented by the straight (slanted) line in the graph, starting at the origin (0 density, 0 encounter) and at a certain angle with the (horizontal) density axis. The slope of the line is the proportionality factor, determined by the level of ease (or difficulty) in finding the prey within the site: the easier it is to find, the steeper the slope. This serves as a reference line for drawing a profitability curve.

Now that the profitability (i.e. the number of prey found, captured, and dealt with per unit time) could never exceed the number found (encountered), the profitability curve should always be below the slanted reference line, except at the origin: no prey, no encounter, no profit. When the prey was present but very scarce, only a very few individuals would be encountered per unit time. Then, the proportion of total time spent in hunting prey would be used up mostly in searching: or conversely, little time would be spent in dealing with the prey captured. Thus, the steepness of the profitability curve would initially be very close to, albeit never steeper than, the reference (encounter) line.

As the prey abundance increased a little further, accordingly a few more individuals would be encountered and captured and, hence, a slightly higher portion of time would be spent dealing with the catches. So, the number found, captured, and brought home per unit time (i.e. the profitability), albeit increasing with prey abundance, deviates from the slanted (encounter) line a little further downwards, and so on. For a further increase in prey density, the profitability curve should sooner or later approach an asymptotic level. This is because, at a high level of prey abundance, comparatively little time would be needed to find prey, whereas dealing with the catches would occupy most of the total time spent hunting. In other words, the profitability curve (the number of prey brought home per unit time) would practically reach its ceiling and would stay there no matter how much more abundant the prey became from then on. [The profitability curve is well described by the popular 'disk equation' of C. S. Holling (1959): details are in Royama (1970), pp. 642–3.]

1.5 Allotment of Hunting Time among Different Sites

I now look into how the tit would allot its time hunting among potential sites. The foregoing profitability curve provides the basis for this inquiry in the following idealization. [Here, I generalize 'profitability' in terms of the total biomass of food fed to the chicks per unit effort (energy expenditure) of hunting.] For simplicity, consider how the tit allots its time between two sites, A and B, and let P_A and P_B designate the level of profitability at A and B, respectively. Further, let P_A vary, while P_B is fixed at its maximum level. Let us consider the three situations: P_A is more or less equal to P_B; P_A is very low; and P_A is in between.

In the first situation, there would be no advantage (no reason) for the tit to select A over B, or vice versa. Then, the allocation of time between the two sites would become a matter of arbitrary choice and, hence, would be subject purely to a chance variation. That is, we would expect a fifty–fifty allocation of time between A and B. Second, if P_A is very low, we would expect that the tit would allot little time to site A. In the third situation, in which P_A is less than P_B but not very low, what we would expect to happen is not readily conceivable.

Theoretically, there would be no point for the tit to allot any portion of time to site A when P_A was less than P_B. But it is not realistic to expect that the tit spends little time in A until P_A becomes close to P_B and that, at that moment, there occurs a sudden jump in time spent at site A. What we would expect is the following. The tit spends a certain amount of time at A to monitor changes in P_A so as not to run the risk of prematurely accepting or rejecting the site as profitable or unprofitable. On the other hand, spending too much time would be a waste. It follows that the tits must optimize the allocation of time. This depends on their ability to assess P_A. There could even be considerable individual variations among the tits under varying environmental conditions. [Note: You might wonder if the tits have this sort of cognitive ability to assess the profitability. Surely it is unlikely that the tits solve the problem intellectually. However, I am not surprised if their instinctive behaviour, evolved through natural selection, would guide them to the solution.]

1.5.1 Consequence of Hunting by Profitability

In these circumstances, what consequence do we expect in terms of the amount of prey (from site A) fed to the chicks in the nest?

Suppose that we made repeated observations to measure the proportion of the total time the tit allotted to site A, and plot it on the vertical (percentage) axis against prey abundance at the site on the horizontal axis. It would most certainly form a scattergram along a sigmoidal curve as an average trend: starting with a slow rate of increase, followed by a rapid increase, but slowing down again to level off. At what level of prey abundance and how rapidly the curve (average trend) starts increasing should depend on the efficacy of the tits' instinctive assessment.

Now, multiplying the profitability P_A of site A by the time allotted to the site, we would expect in principle that the amount of prey (from A) fed to the chicks per unit time, as plotted against the prey abundance (in A), is likely to be sigmoidal. This explains the trend that Luuk Tinbergen actually observed in his study with respect to *Panolis flammea* larvae: few of the larvae were fed to the chicks when scarce in the woods; the proportion of the larvae in the diet increased substantially when moderately abundant in the woods; but, for a higher level of abundance, the percentage in the diet did not keep increasing but levelled off.

1.6 Hunting by Profitability as Principle

My conception of hunting by profitability is not just a means of comprehending the tits' behaviour. As I see it, the survival (and fitness) of an animal would in principle depend on how to allot optimal amounts of time (effort, energy) among its feeding sites (niches) according to the profitability that each site would provide. I will come back to this issue in Chapters 8 and 9 in which I talk about the evolution of niche selection by animals in general.

2 · *The Paradox of Crypsis: Is it Effective against Visual Predation?*

It is generally accepted that the cryptic colouration and mimetic shapes of many insects have evolved as a protective device against visual hunters. I have no problem accepting this in principle. Rather, I am positive that visual predation is the only possible route in the evolution of many types of crypsis. However, I have some problems with the details.

The Dutch biologist L. de Ruiter (1952) performed a laboratory experiment with the so-called stick caterpillars, the larvae of the genera *Ennomos* and *Biston* (Lepidoptera: Geometridae), which so perfectly mimic real twigs. He offered these caterpillars (anaesthetized and motionless), mixed with twigs, to a few young jays (*Garrulus glandarius*) and chaffinches (*Fringilla coelebs*). These birds were hand-raised and had never seen these objects before. Nonetheless, the young birds tended to manipulate the objects with seemingly no particular purpose and, while pecking at them indiscriminately, happened to find that some of them were edible. de Ruiter repeated the experiment by varying the proportion of caterpillars to twigs. When the proportion was high the birds were rewarded frequently enough to be encouraged to keep pecking at the objects. On the other hand, when the proportion was low, the birds tended to lose interest sooner. Thus, de Ruiter concluded that the shape of these caterpillars mimicking twigs was indeed an effective device to protect themselves from bird predation in the field. He further suggested that the mimicry would effectively protect the insects as long as they were dispersed in the habitat, so that the chance of being accidentally picked by birds would be practically nil. In other words, the mimicry actually works and hence has evolved.

Meanwhile in England, Bernard Kettlewell (1958) was investigating the suggestion that the so-called industrial melanism in certain lepidopteran species evolved as a protective device against bird predation. These species are known to have two contrasting morphs: the phenotypes of the adults are either pale or dark, although an intermediate morph does exist. Before the Industrial Revolution, the pale *typical* morph was predominant in Britain. Since then, the dark *melanic* morph has largely replaced the former in areas where the moth's habitat became sooty with industrial pollutants, hence the name 'industrial melanism'. Apparently, the *typical* morph is now coming back where the industrial pollution is easing.

Kettlewell used adults of the peppered moth (*Biston betularia*) for his investigation. He conducted a mark–release–recapture experiment in the woods in two areas: a polluted area near the city of Birmingham and an unpolluted rural area in Dorset. Using a small dab of paint, he marked an equal number of the laboratory-raised adults of both morphs, released them in the woods at dawn, and recaptured them the following night (together with unmarked wild moths, of course). He found that, among the recaptured moths, the proportion of the *melanic* form was substantially higher in the polluted woods and vice versa in the unpolluted area, demonstrating that the morph that matched the background survived better. Because these moths tended to rest on tree trunks during the daytime, he concluded that the cause of the difference was differential predation by birds: the birds must have more readily taken whichever morph was conspicuous against the background. To demonstrate this, Kettlewell conducted experiments with birds, first in an aviary and then in a semi-wild set up. He took films (there were no videos then!) of bird predation in action in the latter set up. He placed several individual moths from each of the *typical* and *melanic* forms on a dark tree trunk and another group on a lighter trunk. A robin (*Erithacus rubecula*) hanging around nearby spotted the moths, flew to them, and took whichever was the more conspicuous form in quick succession. It was impressive to see that the morph that did not match the background was so vulnerable to birds; it was so convincing that bird predation must be a major cause of the *melanic* form becoming more frequent in an industrial area compared with the *typical* form; and it demonstrated beyond doubt that crypsis was working as a protective device.

Nonetheless, I had a problem, or more like a dilemma. As described in Chapter 1, from the late 1950s to the mid 1960s, I photo-recorded the insects that the great tits (*Parus major*) brought to nests for their young over five field seasons: two in Yamanaka, Japan, and three in Wytham Woods near Oxford, UK.

What startled me, or rather confused me, was the following. In Wytham, lepidopteran larvae were the major source of the nestlings' diet. Among those, the larvae of two geometrids, feathered thorn (*Colotois pennaria*) and pale brindled beauty (*Phigaria pilosaria*), and those of the noctuid, green-brindled crescent (*Allophyes oxyacanthae*) were brought to the nests in large numbers. At the peak, more than 150 or even 200 larvae were brought to the nest daily by a pair of parent tits. In fact, these comprised more than half (in frequency) of the nestlings' food from late May to mid-June. Why was I surprised? The larvae of the first two geometrids are stick caterpillars. Albeit belonging to different genera, they both mimic twigs as perfectly as those stick caterpillars that de Ruiter used in his experiment. The noctuid *oxyacanthae* larvae perfectly mimic pieces of bark on the trunks and thick branches of their main food plants in Wytham, the common hawthorn (*Crataegus monogyna*) and the blackthorn (*Prunus spinoza*). All these caterpillars are nocturnal feeders and, during the daytime, each stays still on the place that it mimics. So, if you literally interpret de Ruiter's conclusion that mimicry works and therefore has evolved, you would encounter a conflict: my tits never seemingly had a problem finding them. For example, a pair of tits once collected 16 *oxyacanthae* larvae in 28 minutes; that is, each parent took, on average, one larva every 3.5 minutes. These minutes included: flying out to the hunting site, searching for a larva, capturing it, bringing it home, feeding it to a chick, and flying out to hunt for another. Evidently, each parent bird had only about a minute or even less to find such a cryptic larva in each of the eight quick cycles of hunting trips.

Thus, I had no choice but to conclude that, at least in my observations in a totally wild environment, mimicry did not effectively protect the caterpillars from bird predation as de Ruiter had suggested. Then, I noticed that there was a jump in his interpretation. For one thing, his experiment was carried out in an artificial set up in which the experimental subjects (the young naive birds) were deprived of the chain of clues that they could instinctively have followed to reach the prey in their natural environment. In other words, his experiment and my observation were not directly comparable. Nonetheless, I could not reject the idea that mimicry evolved under pressure from visual predation: I could not think of any other possibilities; hence, the dilemma. After a while, however, I found a way to get around it.

The level of perfection in mimicry that we see today must have evolved a long way from a primitive form that could have been quite different. One day, long ago, a predator began to hunt its prey visually and the game of hide and seek started. A slightly less conspicuous mutant prey had a better chance of survival, but it was, in turn, countered by a slightly more efficient predator, and so on as the game went on. That is, in each cycle of hide and seek, the predator must have been exerting a constant pressure that pushed a primitive form of crypsis towards a more 'effective' and 'sophisticated' form, and so on to the current level of 'perfection'. In other words, in this game, perfection (in the usual sense) would never have been achieved: crypsis, no matter how 'perfect' as it might appear to us, would never be 'perfect' from the point of view of the predator because it, too, would have developed a keener eye. After all, the evolution of crypsis is a continuous process of selection in favour of a slightly better form as against a not-so-good one among the prey, as well as in favour of a keener-eyed individual among the predators. On the other hand, if we assume that crypsis effectively protects the prey against visual predation, we encounter a number of contradictions.

If, during a game of hide and seek, the prey attained a level of crypsis at which the predator could no longer find the prey efficiently, what would happen? I would think that the predator would pay little or no more attention to the prey as it is an unprofitable source of food, as de Ruiter's experiment seems to suggest: too many dead sticks discouraged the hunters. In other words, the predator conceded defeat and the prey won the game. But this contradicts my observation that the tits are still finding the caterpillars with ease, and I cannot see a win–defeat situation likely to happen.

Imagine that predation pressure had pushed the evolution of mimicry to perfection and, all of a sudden, the predator became virtually unable to find them. This would be most unlikely to happen. For one thing, before the level of perfection had been achieved, there must have been an intermediate stage at which the effectiveness of mimicry was not yet so high that the predator could still find them comparatively easily. As a more advanced (sophisticated) form prevailed among the prey so that the mimicry began to work more effectively, predation pressure would have begun to ease off because effective mimicry means unprofitability for the predator. Then, the evolution of mimicry would have slowed down. This process of hide and seek, sooner or later, would promote an equilibrium state. Under these circumstances, the level of mimicry should have remained mediocre, comparatively speaking. Then, for the prey to achieve the current level of sophistication at which, as de Ruiter suggested, the prey is effectively protected from predation, the evolution of mimicry must have kept going by itself; it must have kept going even after predation pressure had virtually been removed. This would be very unlikely to happen: it sounds more like a Lamarckian sort of evolution rather than Darwinian.

Another argument could be that the current level of perfection in mimicry was already achieved long ago and since then the predators have developed a method to detect the prey other than a visual means. So, the current form of mimicry has remained neutral. But this view contradicts the Kettlewell observation: the birds are actually using visual cues as a primary means to detect the more conspicuous morph. Neither could the argument answer the question: when and how was that level of perfection achieved before the bird developed the method of detection other than the visual cue? Then, we would have to go back to the question: how had mimicry achieved the level of 'perfection' in the first place? And so goes the circular argument.

You might, once more, argue that mimicry is effective against naive young birds and so evolves. But this does not make much sense either. Would it be possible for a given genetic trait to prevail under a weak selection pressure while, simultaneously, being exposed to a heavy pressure exerted by experienced (adult) predators? I would think that those prey individuals that had escaped the weak pressure would have perished under the heavy pressure. In other words, a weak pressure in the simultaneous presence of a heavier pressure would be an unlikely force of selection. Incidentally, by the time the fledglings (of most insectivorous birds) are out of the nest and begin to fend for themselves by hunting prey visually, those early-season caterpillars will have long pupated under the duff: they would have rarely been exposed to predation by the naive young birds.

As far as I am concerned, there is only one way to resolve these contradictions, and that is to recognize the paradox that mimicry is not an effective means (in the absolute sense) of protection against visual predation and, therefore, has evolved. The evolution of mimicry must have been a gradual and continuous process, such that the higher the selection pressure the higher the level of perfection would become. From this point of view, it makes good sense that the larvae in my observations that achieved an apparently high level of mimicry (*sensu* de Ruiter) are still under heavy predation by birds, the visual hunters. The game must be ongoing with no clear winner.

Don't take my conjecture as definite. It must be tested and should be debated, needless to say.

3 · *Logistic Law of Population Growth: What Is It Really?*

3.1 Preamble

In 1973, the Government of Canada, my employer, somehow found a surplus in its annual budget that could be spent on science research, a very unusual occasion. We, the research scientists, were encouraged to apply for 'educational leave', i.e. to go to an appropriate institution abroad to learn the latest development in the area of his/her work. Around that time I had begun to work on the analysis of spruce budworm outbreak processes in eastern Canada, and I wanted to learn about stochastic processes, a branch of probability and statistics that deals with time-dependent processes. So I wrote to Professor Patrick (Pat) A. P. Moran of the Australian National University, asking if I could study under his guidance. I had known him through his mathematical work on population ecology and genetics (well-known among ecologists for his statistical analysis of the Canada lynx ten-year cycles). He said I would be very welcome. So, in May 1974, I headed Down Under to spend a full year in Canberra with my family (my wife and an infant son) with all expenses for travel, etc., on the government tab.

One morning, after a tutorial session in his office, Pat took out a reprint, handed it over to me and said: 'Tom, you might be interested in reading this'. It was an article by William (Willy) Feller (the author of the highly regarded two-volume textbook: *An Introduction to Probability Theory and its Applications*). The article was 'On the logistic law of growth and its empirical verifications in biology' (Feller, 1940). In it, the author pointed out that it was a law without a theory and that it would be dangerous to rely solely on empirical verifications. Pat gave it to me, hinting that I might think about a possible theory. Although I kept the idea in mind, I shelved it for several years. One day, I was playing with dots and circles on a sheet of graph paper to figure out a problem about the geometry of a predator searching for prey. Suddenly a thought struck and I 'slapped my lap', a Japanese expression for an 'aha!' moment. For the next few weeks, I concentrated on the subject and managed to derive the logistic equation from the ecological principle of intraspecific (within-population) competition, originally published in Royama (1992, section 4.2.5). Since then, the theory I conceived has become a fundamental basis of my ecological studies of populations that I talk about in this book. In the following, I begin with a short history of the development of the classical logistic model.

3.2 The Classical Logistic Equation

Back in mid-nineteenth-century Belgium, the mathematician Pierre-François Verhulst, inspired by his mentor Adolphe Quetelet during a discussion on socioeconomic issues, began to work on the problem of the then influential idea of the Malthusian geometric progression of populations. [Thomas Robert Malthus was the English cleric and the author of the well-known (1798) treatise 'An Essay on the Principle of Population'.] The Malthusian theory (as it is popularly known) is essentially this: a human population grows at a rate of geometric progression, whereas essential resources (food supplies) grow only at a rate of arithmetic progression. Therefore, unless the population was controlled by some social or moral disciplines, only catastrophes like epidemics, famines, or wars could reduce the population to a level at which the resources could sustain it. [A geometric progression is a series of numbers of offspring such as: starting with a founding pair, increasing like {2, 4, 8, 16, 32, ... and so on} to sooner or later 'explode' without bounds. In contrast, an arithmetic progression goes steadily like {1, 2, 3, 4, ...}.]

As an antithesis of the Malthusian perception, Verhulst (1838) reasoned that a human population would not increase at a geometric rate because of the limited productivity of resources; therefore, it would not increase explosively even before it reached a food limit or was accidentally hit by an epidemic, etc. The resultant mathematical model is the classical Verhulst differential equation that generates the well-known sigmoid growth curve, which he called 'logistique' in his later article (Verhulst, 1845) as opposed to 'logarithmique' (by which he meant the Malthusian geometric progression).

However, Verhulst's idea of logistic growth of population did not attract much attention from the people of his time and had long been forgotten until its apparent and inconspicuous rediscovery around 1910.

3.2 The Classical Logistic Equation

However, its significance was not recognized among ecologists for another decade until Raymond Pearl (Professor in Biometry and Vital Statistics, Johns Hopkins University) with the collaboration of his fellow biometrician Lowell Reed (later the president of the university) presented a paper (Pearl and Reed, 1920) in which they proposed an equation that happened to be virtually (if not exactly) the same form as Verhulst's original. [Apparently, Pearl and Reed did not know the earlier work of Verhulst, so they did not use the term 'logistic'.] Pearl and Reed demonstrated that their equation fitted the census data of the United States since 1790 almost perfectly: I talk about its details in Chapter 4.

Thus, there are two versions of the classical logistic equation in the form of a differential equation. However, these two originals are seldom cited or used (in their original forms) in current ecology. Instead, many standard textbooks employ a third (or 'common', as I call it) version. So, there are three versions altogether:

$$dx/dt = (\alpha - \beta x)x \qquad \text{Verhulst version} \qquad (3.1a)$$

$$dx/dt = a[(b - cx)/b]x \qquad \text{Pearl–Reed version} \qquad (3.1b)$$

$$dx/dt = \rho(1 - x/K)x \qquad \text{Common version} \qquad (3.1c)$$

where x stands for population size (density), t for time, and the other alphabetical letters for constant parameters: some notations and expressions are mine for convenience of writing without changing the original forms.

We see that these are so similar to each other in mathematical form that the common version (3.1c) is often referred to as the Verhulst–Pearl logistic equation. Notice that version (3.1c) is identical in every aspect to version (3.1b), whereas it differs from version (3.1a). In fact, (3.1b) and (3.1c) contain certain undesirable attributes, as I will reveal in due course.

Notice further that these versions are similar because all of them are exactly of the form $dx/dt = xf(x)$ in which $f(x)$ is a linear and decreasing function of x, i.e. $f(x)$ when plotted against x exhibits a straight line with a negative slope. But, then, what is the ecological significance of the form that the equations in (3.1) assume? To understand it, we must recognize the simple but fundamental nature of animal population processes.

3.3 Fundamental Nature of Population Processes

Every animal population, be it an insect, fish, bird, or even human population, changes as a balance between gain and loss, and nothing else. There are four processes that determine the gain and loss: birth, death, immigration, and emigration, and nothing else. These processes are determined by interactions between the organism and its environments, physical and biotic. The effect of these processes on population changes succinctly manifests in the rate of change in population through the course of time. In fact, it is the essential point in studying population dynamics to look at, to think about, and to understand the time process of the rate of change. From this point of view, let us look into the ecological significance of the differential equation of the form $dx/dt = xf(x)$.

3.4 Ecological Significance of the Differential Equation: $dx/dt = xf(x)$

Let us consider a short period (interval) of time and designate its length by Δt. [The symbol Δ (the upper-case Greek letter 'delta') is often used for denoting a small amount of 'change', 'difference', 'interval', 'increment', etc., without precisely specifying its magnitude for the moment.] Correspondingly, there is a small change in the population, denoted by Δx. That is, over the interval Δt, the population changes from x to $(x + \Delta x)$. Then, the rate of change in the population over the interval is the ratio $\Delta x/\Delta t$ because it is equal to the difference $[(x + \Delta x) - x = \Delta x]$ over the time period (interval) Δt. This is a *per-unit time rate of change* in the population as a whole, or the *gross rate of change*. But, as will become clear in due course, it is a little more insightful to consider an average rate per individual of the population, or a *per-capita rate of change*, given by the gross rate of change divided by the population size $x > 0$, i.e. $(\Delta x/\Delta t)/x$.

Now notice that the rate of change $(\Delta x/\Delta t)$ serves as an index of the direction of a change in population. Consider that the population x increases to $(x + \Delta x)$ over Δt. Then Δx is an increment, i.e. a positive quantity, such that the rate $(\Delta x/\Delta t)$ is positive. If x is decreasing, then Δx is a decrement, i.e. a negative quantity, such that the rate $(\Delta x/\Delta t)$ is negative. And $\Delta x = 0$ when no change occurs in x over Δt, i.e. $(\Delta x/\Delta t) = 0$.

Thus, for a natural population to grow from low to high levels but not to exceed a certain level, its rate of change $(\Delta x/\Delta t)$ must be positive when x is low but must decrease to zero when x approaches a

3.4 Significance of the Differential Equation

certain level. Incidentally, what if the population happened to exceed this certain level? The rate ($\Delta x/\Delta t$) must become negative so that x decreases.

All in all, for the population to be prevented from obeying the Malthusian progression, the per-capita rate of change ($\Delta x/\Delta t$)/x has to be changed according to changes in x. That is, the rate of change ought to be a function of x, i.e. written in general form:

$$(\Delta x/\Delta t)/x = f(x) \tag{3.2}$$

in which the function $f(x)$ must satisfy the following prerequisites: positive when x is low; negative when x exceeds a certain level; and 0 at this certain level.

Now, in order to see what the (per-capita) rate of change (3.2) really means (in terms of, say, a curve as a graphical realization of the values of x as plotted against t), we have to specify what these Δt and Δx actually are. This depends on what sort of a population process we have in mind. There are two major classes of population processes: one is continuous in time, and the other discrete.

In a continuous-time process, birth and death take place at any point in time during the population process, like a human population or the fruit fly (*Drosophila*) population in a milk bottle; for now, let us ignore migration for simplicity. In this class of process there is no clear separation among generations; at any point in time, the population is made up of many different age groups; each individual has its own schedule of reproduction, so to speak. To simplify the situation for an easy mathematical treatment, let us assume that the population changes only in its size x as time goes by, ignoring its age structure. Thus, it can be assumed that the process of population change at any interval of time obeys the same rule, as stipulated by (3.2), at any arbitrary point in time over the entire period of population growth under consideration.

In a discrete-time process, reproduction in the population occurs at a certain distinct point in time, like the breeding season in birds and insects in a temperate zone. Thus, the population size x increases only during the breeding season and, outside the season, it only decreases due to mortality. In other words, the manner in which x changes from time to time cannot be the same all the time. Only when we consider a *net* change between two consecutive generations, observed at a certain developmental stadium (e.g. egg-to-egg, larva-to-larva, or pupa-to-pupa rate of change), can it be assumed that the process obeys the same rule.

Verhulst, as well as Pearl and Reed, was primarily interested in human populations and, therefore, in a continuous-time process.

3.4.1 Continuous-Time Process

Consider that Δt in the rate (3.2) is extremely short, i.e. an instant. Then, the corresponding Δx is an extremely small increment or decrement (infinitesimal): keep in mind that Δt is not exactly 0. Then, the ratio $\Delta x/\Delta t$ is the rate of change in the population at the instant moment t, i.e. the instantaneous rate of change. In calculus, Δt as an instant moment in time is written dt and the corresponding infinitesimal increment Δx as dx, and the expression dt or dx is called a 'differential', and the instantaneous rate dx/dt is called the 'derivative' (of x with respect to t). Replacing Δx and Δt in equation (3.2) with the differentials dx and dt, respectively, we have the differential equation $(dx/dt)/x = f(x)$. If we equate $f(x)$ to the expression $\rho(1 - x/K)$, we have the common-version logistic equation (3.1c). If $f(x) = (\alpha - \beta x)$, we have the original Verhulst equation (3.1a). But, how did Verhulst (1838) conceive his version (3.1a) and Pearl and Reed (1920) theirs (3.1b)? [Note: As already mentioned, (3.1b) is identical in form with the common version (3.1c) because $a = \rho$ and $c/b = K$.] The way Verhulst conceived his is essentially as follows.

He began his argument with the form $dx/dt = xf(x)$, but did not suggest any particular model for $f(x)$: this is presumably because he pioneered the subject and no previous study preceding his was available. So, instead, he represented the quantity $xf(x)$ in the form of a general polynomial function in x. He argued that the first-degree polynomial (i.e. $dx/dt = \alpha x$) would yield an unbounded (Malthusian) progression in x. Thus, to make the population bounded from above, he suggested that a higher-degree polynomial be required, the simplest being the second-degree, i.e. $dx/dt = \alpha x - \beta x^2$, which is his original logistic equation, i.e. the version (3.1a). He said he tried the polynomials up to the fourth degree but found the second degree was good enough to describe some sets of actual demographic data from France, Belgium, and England. Thus, no ecological meaning is attached to the parameters α and β.

As for the Pearl–Reed version (3.1b), the authors did not derive it from scratch. Instead, they conceived their equation by analogy with a chemical process known as autocatalysis. [Note: I critically examine their analogy in Chapter 4.] Then, upon differentiating it, they obtained their equation (3.1b). Therefore, again, no concrete ecological meaning is attached to the parameters (a, b, or c).

3.4 Significance of the Differential Equation

The common version (3.1c) is an algebraically compact form of (3.1b) in which, as already shown, $\rho = a$ and $K = b/c$: there is no difference in mathematical attributes. I do not know (to my ignorance) who invented the common version, but it can be derived in the following manner. Consider that an actual population has an upper limit, say K, such that $0 < x \leq K$. [Note: $x = 0$ should be excluded because nothing would grow from nothing.] Dividing each term of the inequality by K, we have $0 < x/K \leq 1$. Then, transferring x/K from the middle to the right, we have: $0 \leq 1 - x/K$. Clearly, the expression $(1 - x/K)$ has the attribute that when x is very small to start, it is close to 1; as x increases to approach K, the expression accordingly decreases ever more closely to 0. So, if the per-capita rate of change $(dx/dt)/x$ is proportional to $(1 - x/K)$ with proportionality factor ρ, the population should increase (from below K) all the time, but with a diminishing rate, and should eventually level off at K. Thus, we have the common version (3.1c).

Now, these differential equations in (3.1) can be solved to evaluate the respective equations in the form $x = f(t)$, i.e. x as a function of t. [Note: I use the symbol f for denoting a generic function. In concrete forms, the f here may differ from the f in (3.2).] Appendix 3A shows how to solve the differential equations (3.1c and 3.1a). The resultant logistic equations, corresponding to (3.1a) and (3.1c), respectively, are given by:

$$x = x_o e^{at} / [1 + (e^{at} - 1)(\beta/\alpha)x_o] \quad \text{Verhulst version} \tag{3.3a}$$

$$x = x_o e^{\rho t} / [1 + (e^{\rho t} - 1)(1/K)x_o] \quad \text{Common(textbook) version} \tag{3.3c}$$

in which x_o is an arbitrary initial population size, and e is the constant number equal to 2.718281828459..., commonly known as the base of the natural logarithm. As you see, the right-hand side of (3.3a) or (3.3c) is a function of t, i.e. $f(t)$, and hence we have evaluated $x = f(t)$. By assigning certain *appropriate* values to the parameter sets (α, β) and (ρ, K), and starting from an arbitrarily chosen initial (small) population size x_o, each of the equations (3.3a and 3.3c) generates the familiar sigmoid growth curve.

The fact that the Pearl–Reed equation – usually given in the common form (3.3c) – describes so well not only the US census data but also the populations of many other species has drawn the considerable attention of ecologists. However, few people could find ecologically probable mechanisms underlying these logistic models because the way the original authors conceived the models did not provide a key. So, some ecologists have tried to make ecological sense of the models, especially of the parameters ρ and K of the common form.

3.4.2 Conventional Interpretation of the Logistic Equation

First of all, the expression $(1 - x/K)$ on the right-hand side of version (3.1c) appears to fit the perception of Pearl and Reed that the rate of population increase at any instant of time is proportional to 'the still unutilized potentialities of population support existing in the limited area'. Ever since then in ecology, the parameter K has been interpreted as the 'carrying capacity' of the environment. In particular: when x is very low, the environment is almost 100% 'vacant'; as x approaches K, the vacancy decreases ever more closely to 0%; and in between, the population grows at a rate proportional to the vacant (still unutilized) portion.

Royal Chapman, in his then popular textbook *Animal Ecology* (1931), envisaged that the limited growth rate was due to the resistance of the environment and coined the term 'environmental resistance', an idea similar to Adolphe Quetelet's 'socio-economic force', some 80 years earlier.

Now, when x is very low and the environment is nearly 100% vacant, i.e. $(1 - x/K)$ is almost 1, a realized (per-capita) rate of increase $(dx/dt)/x$ in model (3.1c) is practically at its maximum which, in fact, is equal to the proportionality constant ρ. In other words, ρ is the maximum possible rate of increase that would be realized when there is little or no 'resistance' from the limited capacity of the environment. Hence, in conventional ecology, ρ is called 'the innate capacity for increase' or 'intrinsic (potential) rate of (natural) increase'. [Incidentally, the parameter ρ is usually designated by the letter r but, because I use that in another context, I use ρ here instead.]

At first, these perceptions sound reasonable, as they are intuitively appealing interpretations. However, if you think about them a little more carefully, rather than memorizing them in order to get a pass mark in a term exam, you would find these perceptions to be rather vague. In fact, these are interpretations through metaphors in an effort to make ecological sense of the equation only after it had been given in the Pearl–Reed (1920) article. And the equation was conceived (by analogy with a chemical reaction) primarily as a mathematical device to tame the wild Malthusian population.

The question remains unanswered: what are these interpretations in terms of ecological factors, conditions, or processes that we can actually observe in nature or in the laboratory? In other words, behind these ideas there is no ecological necessity that the equation should be what it is. Thus, as the mathematician Willy Feller (1940) pointed out, the pattern of population growth as depicted by equation (3.1c) or its solution (3.3c) is an empirical law without theory, or it is a mathematical model without much ecological bearing. In other words, it is a good descriptive model but provides little insight into the ecological mechanism actually underlying the observed population growth patterns.

Nonetheless, I was convinced of the numerous demonstrations since Pearl and Reed that had shown so many animal populations from diverse taxa obeying this law: be it a *Drosophila* population in a milk bottle or the human population in America. I was convinced that there must be ecological necessities. To uncover them, I started thinking in a discrete-time process, rather than in a continuous-time process, as I was more familiar with forest birds and insects.

3.5 Discrete-Time Processes

There are two types of discrete process. One is represented by plotting the continuous logistic curve at a regular interval in time: the process under observation is continuous in time but the population changes are recorded (observed) in discrete intervals in time. Nothing is new here as to the actual mechanisms involved. So, this is not interesting in the present context. The other type of process is truly discrete in time as in many insect and bird populations with distinct breeding seasons. For simplicity, let us consider the population of a univoltine insect in which generations do not overlap. That is, all adults die after laying eggs at the end of the generation and the new generation follows to repeat the cycle: no migration is considered. Also, let us assume that the sex ratio is stable from generation to generation; make it simply 1:1 without much loss of generality.

Consider that we measure the density (x) of individuals at a given stage in each generation: egg, larva, pupa, adult, or whichever is convenient for a quantitative observation. Thus, we have the discrete sequence $\{x_1, x_2, x_3, \ldots, x_t, \ldots\}$ in which the subscript t represents the t-th generation. Then, the net rate of change in population from generation t to $t+1$ (observed at the given stage in each generation), or the intergeneration (net) rate of change, is measured by the ratio x_{t+1}/x_t, written compactly as r_t.

40 · Logistic Law of Population Growth

[Note: Precisely speaking, r_t is the 'per-capita net rate of change in population'. However, it is synonymous with the 'mean reproductive rate of an individual' with respect to the selected stage. So, I may use the two expressions interchangeably. Also, I may drop the adjective 'net' or 'per-capita' if too repetitive. Furthermore, it should be noted that if we select a different stage, the rate r_t would generally differ in value, but the mechanism that determines the rate is the same in principle, given the life cycle. So, we need not specify what stage we select in the present context of theoretical investigation.]

In the following, I consider how $r_t = x_{t+1}/x_t$ can be ecologically determined in the above idealized situation. At first, I build a geometric model (diagram) of competition among the individuals of a population.

3.5.1 Geometric Model of Intraspecific Competition in a Discrete-Time Process

To begin, I suggest that we pick up a sheet of graph paper and assume that it represents the 'effective' habitat space of the organism concerned. Place (mark with a pencil) a number of dots haphazardly over the paper and assume that each dot represents an individual organism. Draw a circle with the radius of a fixed length around each dot at the centre. Name this diagram Figure 3.1.

Now, assume that each circle represents the area within which the individual (the dot at the centre) finds the minimum sufficient amount of resources to survive and reproduce normally. Let the area of the circle be

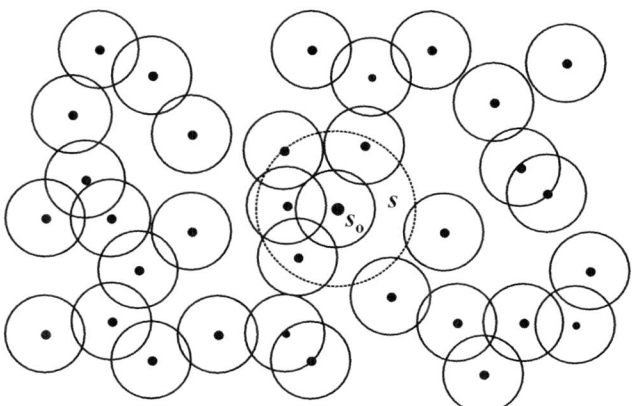

Figure 3.1 Geometric representation of competition.

s_o: that is, s_o is a measure of the minimum sufficient requirement of resources by an individual (a dot) given the amount of resource supply in the habitat. Assume further that survival and reproduction of the dot would be adversely affected if the resources fell below a certain level, whereas a surplus has no influence.

In the diagram, the circle of a given dot may or may not overlap the circles of other dots. If it does, the dot at the centre is assumed to compete with the other dots over the shared portion of the resources. We see then that the number of these competitors is equal to the number of the dots within the concentric (larger, dashed) circle with the radius that is twice as long as those of the other smaller circles. Let s be the area of this larger circle, i.e. $s = 4s_o$. [Note: In the following, I may call s '(individual) resource requirement' although it is four times larger.]

We now move on to formulate a mathematical model of population growth with which we can evaluate the reproductive rate $r_t = x_{t+1}/x_t$.

3.5.2 Formulation of the Model of Population Growth

Although this section is mathematically involved, I suggest that you go through it step by step, rather than just glance through it, because it is very important to understand what the model is all about. In fact, the discrete-time model I develop here serves as a basis of all population process models that I develop and use in the subsequent chapters. So let's go.

To begin, consider the process within a given (say t-th) generation. Assume in particular that, given the environmental conditions (e.g. resource supplies), the average reproductive rate of an individual depends on the number of competitors. Let $r_{(i)}$ stand for the mean reproductive rate of those individuals in the population that have i competitors within each own circle of area s ($= 4s_o$) as in Figure 3.1. [Note: The notation $r_{(i)}$ is a reproductive rate of an individual (when it has i competitors within the t-th generation) to be distinguished from the intergeneration rate of change r_t ($= x_{t+1}/x_t$).] Also, let $\Pr(i)$ be the expected proportion (or the probability of occurrence) of these individuals that have i competitors. Furthermore, assume that the resource supply is sufficient in all generations such that the size of area s_o in Figure 3.1 remains unaffected, i.e. invariant between generations: that is to say, the supply may be depleted within a generation but will be recovered by the beginning of the following generation.

Under the above circumstances, the intergeneration rate of change r_t is equal to the mean reproductive rate in the t-th generation, which in turn is given by the sum of $r_{(i)}$, weighted by its frequency occurrence $\Pr(i)$, for all $i = 0, 1, 2, \ldots, \infty$ (infinity), i.e.

$$r_t = r_{(0)}\Pr(0) + r_{(1)}\Pr(1) + r_{(2)}\Pr(2) + \ldots + r_{(i)}\Pr(i) + \ldots \quad (3.4)$$

Now, to evaluate r_t, we need to specify $r_{(i)}$ and $\Pr(i)$ on the right-hand side. To do this, let us assume the following.

First, each individual has a potential (maximum) reproductive rate that is realized when there is no competitor, i.e. when $i = 0$. Thus, $r_{(0)}$ is the mean potential reproductive rate of an individual in the population, or simply the potential reproductive rate. Assume further that each time an additional competitor is involved, the potential rate decreases by factor k, a positive constant less than 1. Thus, if a given individual has one competitor, its reproductive rate $r_{(1)}$ is equal to $r_{(0)}k$; if two competitors, $r_{(2)} = r_{(1)}k = (r_{(0)}k)k = r_{(0)}k^2$; and if in general i competitors:

$$r_{(i)} = r_{(i-1)}k = r_{(i-2)}k^2 = \ldots = r_{(1)}k^{i-1} = r_{(0)}k^i, \qquad 0 < k < 1.$$

We see that, starting with $r_{(0)}$, the rate $r_{(i)}$ decreases in a geometric rate as i increases because $0 < k < 1$; the lower the k value, the faster the decrease. In other words, k is a measure of the intensity of competition (or impact on the reproductive rate of an individual): the lower the value of k, the greater the intensity or the impact. [Note: k here is the lower-case letter, not to be confused with the upper-case K of the (common-version) logistic equation (3.1c).]

Substituting $r_{(0)}k^i$ for $r_{(i)}$ on the right-hand side of (3.4), we have:

$$r_t = r_{(0)}\{\Pr(0) + k\Pr(1) + k^2\Pr(2) + \ldots + k^i\Pr(i) + \ldots\}. \quad (3.5)$$

Now that the $\Pr(i)$ is a probability distribution function, it is determined when we specify the spatial distribution of the dots in Figure 3.1. In the present chapter, I assume a purely random distribution, i.e. the location of every dot is completely independent of the location of the other dots. So, assume that we pick up a sheet of graph paper.

Let x be the average number of dots per unit area of the paper, i.e. density. [Note: As we are dealing with a within-generation process, I drop the generation subscript t for now.] Consider that, if we place a circle of size s at random without reference to the locations of dots over the paper, the average number of dots within the circle would be expected to be sx. Then, a circle of area s around each dot (in Figure 3.1) would also contain, on average, sx other dots, excluding the one at the centre. This is because drawing the circle around each dot, as it is distributed completely at random, is equivalent to placing the circle at random without reference to where the dots are. Then, the expected proportion Pr(i) (of these dots, each of which has i competitors) would obey the so-called Poisson distribution (named after the French mathematician Siméon Poisson who conceived it in the early 1800s). Its probability distribution function is given by the formula $(sx)^i \, e^{-sx}/i!$, i.e.

$$\Pr(i) = (sx)^i e^{-sx}/i! \tag{3.6}$$

in which the number e (= 2.71828...) is the base of the natural logarithm. The symbol '!', called 'factorial' in mathematics (not as in 'Wow!' in daily language), is so defined that

$$i! = i \times (i-1) \times (i-2) \times \ldots \times 3 \times 2 \times 1,$$

the special case for $i = 0$ (i.e. 0!) being defined as 1. [Note: I do not show how to derive the Poisson distribution (3.6) here. However, I derive its general version called the 'negative binomial distribution' in Chapter 5. So, wait until then.] Here, I show a numerical example of the Poisson distribution (3.6).

Assuming, for example, $sx = 3$ and $i = 2$, we find in the formula (3.6): $3^2 = 9$, $e^{-3} = 1/(2.71828...)^3 = 0.0497871...$, and $2! = 2 \times 1 = 2$, so that Pr(2) = $9 \times 0.049787 \div 2 = 0.224$. That is, 22.4% of all dots in the habitat space are expected to have 2 competitors. Likewise, the expected proportions (%) for (the number of competitors) $i = 0$ to 10 (for $sx = 3$) are found to be:

i:	0	1	2	3	4	5	6	7	8	9	10
$100 \times \Pr(i)$:	5.0	14.9	22.4	22.4	16.8	10.1	5.0	2.2	0.8	0.3	0.1

That is, if you place 100 dots completely at random on a sheet of graph paper with area (10 cm × 10 cm), i.e. $x = 1$, and draw a circle of the area $s = 3$ cm^2 (or radius ≈ 0.9772 cm) around each dot, you would expect to find, on average: 5 dots without a competitor; 15 dots with one competitor; 45 dots with 2 or 3 competitors; and so on to find practically no dot with more than 10 competitors.

[Suggestion: You might be interested in creating an example of the Poisson distribution of 100 dots yourself on a sheet of graph paper. If so, do the following. Generate (using your computer) a series of 100 independent random numbers uniformly distributed in the interval from, say, 0 to 100; the uniform distribution is one in which the probability of the occurrence of any given number u ($0 \leq u \leq 100$) is equal throughout the interval. Call this series a u-series. Likewise, generate another series as a v-series, and pair it with the u-series. Each pair of numbers gives the location of a dot on the (u,v) coordinate plane, your sheet of paper. Using this plane, you can empirically verify if Monsieur Poisson, or my humble calculation above, was right. Mind you, this is a random process and the observed values do not exactly match their own expected values. Only when you make a large number of trials would you see that the average of observations tends to converge (average out) to the respective expectation given by (3.6), an attribute known as the law of large numbers.]

So much for the within-generation process. I now move on to the formulation of an intergeneration population process.

Substituting the right-hand side of equation (3.6) for Pr(i) on the same side of equation (3.5), and resuming the generation subscript t to r and x (i.e. $x_{t+1}/x_t = r_t$), we find:

$$x_{t+1}/x_t = r_t$$
$$= r_{(0)} \left[1 + ksx_t + (ksx_t)^2/2! + \ldots + (ksx_t)^i/i! + \ldots \right] \exp(-sx_t). \tag{3.7}$$

[Note: The expression '$\exp(-sx_t)$' on the right-hand side reads: e raised to the power of ($-sx_t$). That is, this is equivalent to writing the more usual expression e^u as $\exp(u)$. This expression is just for convenience in writing if u is made up of several different characters. There is no change in the mathematical meaning, and I will use whichever expression is convenient, depending on the context.]

It so happens that the sum of the polynomial terms in the brackets on the right-hand side of equation (3.7) has a very nice feature. In the early eighteenth century, the English mathematician Brook Taylor found an important theorem: if a given function, say $f(u)$, is differentiable an infinite number of times at a given value of u ($= a$, say), then the function can be expanded as the sum of an infinite number of polynomial terms. Particularly useful is a special case in which $a = 0$ called the Maclaurin series, named after the Scottish mathematician Colin Maclaurin. The series expansion is given as follows.

The derivative of the function $f(u)$ (with respect to u) is written $df(u)/du$, which can be more compactly written as $f'(u)$. Further differentiating this derivative gives the second derivative, written $df'(u)/du = f''(u)$, and so on. In general, the ith derivative is denoted by $f^{(i)}(u)$. [Note: the letter i

3.5 Discrete-Time Processes

here is a generic expression to denote an ordered integer, like: 0 or 1 or 2 or .... Although I used i in denoting the number of competitors, here I am using it to represent the order of derivatives.] Then, the Taylor–Maclaurin theorem shows that the following would hold:

$$f(u) = f(0) + f'(0)u + f''(0)u^2/2! + \ldots + f^{(i)}(0)u^i/i! + \ldots \quad (3.8)$$

where $f^{(i)}(0)$ is the ith derivative $f^{(i)}(u)$ at $u = 0$ and is a constant, given $i = 0, 1, 2, 3, \ldots$, as in the following example.

Consider that $f(u)$ is the exponential function e^u, i.e. $f(u) = e^u$, the derivative of which is itself, i.e. $f'(u) = e^u$, and so are all higher-order derivatives, i.e. $f^{(i)}(u) = e^u$ for all $i = 1, 2, 3, \ldots$, as explained in Appendix 3B. Thus, $f^{(i)}(0) = e^0 = 1$ for all i. Then, substituting 1 for every f on the right-hand side of equation (3.8), we find:

$$e^u = 1 + u + u^2/2! + \ldots + u^i/i! + \ldots$$

Substituting the expression (ksx_t) for u in the above relationship, we have:

$$\exp(ksx_t) = 1 + ksx_t + (ksx_t)^2/2! + \ldots + (ksx_t)^i/i! + \ldots \quad (3.9)$$

We see that the sum in the brackets on the right-hand side of equation (3.7) is identical to the sum on the right-hand side of (3.9) and, therefore, is equal to $\exp(ksx_t)$ on the left. Hence, by the rule $e^m e^n = e^{(m+n)}$ and after simple algebra, we find that the right-hand side of equation (3.7) is reduced to:

$$r_{(0)} \exp(ksx_t) \exp(-sx_t) = r_{(0)} \exp(-sx_t + ksx_t)$$
$$= r_{(0)} \exp[-s(1-k)x_t].$$

Substituting the above result in (3.7), we find the net rate of change in population from generation t to $(t+1)$ to be:

$$r_t = r_{(0)} \exp[-s(1-k)x_t].$$

This is the model of population growth in the discrete-time scheme I have attempted to find. At this moment, I change the notation $r_{(0)}$ to r_m, in which the non-italicized subscript $_\mathrm{m}$ is a marker rather than a number, to indicate the potential (maximum) reproductive rate of an individual which is realized when the individual has no competitors. Thus, the model we have wanted to formulate is given by:

$$r_t = r_\mathrm{m} \exp[-s(1-k)x_t]. \quad (3.10)$$

I now show that the original Verhulst version of the logistic model, i.e. (3.1a), is a continuous-time case of the discrete-time model (3.10). The significance of the link between (3.1a) and (3.10) is that the set of parameters (α, β) in (3.1a) can be understood in terms of the ecologically defined parameters (r_m, s, k) of (3.10). In other words, the link provides a theory with which we can interpret the classical logistic law, or my answer to the late Professor Willy Feller whose criticism of the classic had motivated me to investigate the subject.

3.6 Classical Logistic Model as Particular Case of Model (3.10)

First, let me transform model (3.10) into its (natural) logarithm, denoted by 'ln'. Then, by the rule $\ln(mn) = \ln(m) + \ln(n)$, and noting that $\ln(e^{-u})$ $= -u\ln(e) = -u$, because $\ln(e) = 1$ by definition, the right-hand side of (3.10) is transformed to:

$$\ln(r_m) + \ln\{\exp[-s(1-k)x_t]\} = \ln(r_m) - s(1-k)x_t.$$

Also, by the rule $\ln(m/n) = \ln(m) - \ln(n)$, the logarithm of the reproductive rate $r_t = x_{t+1}/x_t$ on the left-hand side of (3.10) is equal to the difference $\ln(x_{t+1}) - \ln(x_t)$. Thus, (3.10) is transformed to:

$$\ln(x_{t+1}) - \ln(x_t) = \ln(r_m) - s(1-k)x_t. \quad (3.11)$$

In the above situation, the population size changes from x_t to x_{t+1} over a (discrete) unit interval of time, because the interval $(t+1) - t = 1$. But, let us consider a more general situation in which the population changes from x to $(x + \Delta x)$ over the (continuous) time interval Δt. To transform the relationship (3.11) into the new situation, we write x_t as x and x_{t+1} as $(x + \Delta x)$. Then, we replace the (ln-transformed) rate of change $[\ln(x_{t+1}) - \ln(x_t)]$ over the unit interval of time with the expression $[\ln(x + \Delta x) - \ln(x)]/\Delta t$, i.e. the ln-rate of change over the interval Δt. Thus, (3.11) is transformed to the new situation:

$$[\ln(x + \Delta x) - \ln(x)]/\Delta t = \ln(r_m) - s(1-k)x. \quad (3.12)$$

I now let Δt become infinitesimal, i.e. it becomes the differential dt. Accordingly, the difference $[\ln(x + \Delta x) - \ln(x)]$ becomes the differential $d(\ln x)$ by the definition of a differential, e.g. $df(x) = f(x + \Delta x) - f(x)$ in the limit ($\Delta x \to 0$): cf. Appendix 3B. In the meantime, by the well-known identity $d(\ln x) \equiv dx/x$ (see Appendix 3C), the left-hand side of (3.12) becomes $d(\ln x)/dt = (dx/dt)/x$ in the limit ($\Delta t, \Delta x \to 0$). [Note:

The symbol '≡' reads 'identical', 'equivalent to' or 'defined as', and the like: that is, '≡' is a special (stronger) case of the usual '=' sign.] Thus, we have the differential equation:

$$(dx/dt)/x = \ln(r_m) - s(1-k)x. \quad (3.13)$$

This completes the mathematical transformation of the discrete-time model (3.10) into the continuous-time process. But we need also to transform one ecological condition on which (3.10) is based.

Recall that, when setting up the discrete-time model (3.4) as the prototype of the model (3.10), it was assumed that the resource supply was invariant (that is, the parameter s was invariant) between generations. Translating this prerequisite into the continuous-time situation means that the resources are continually maintained at a sufficient level. This is usually true in a modern human society in which food is continually produced. In the case of a *Drosophila* population in a milk bottle, food must be continually replenished. This completes the ecological transformation of (3.10) into (3.13). [Note: What if the food supply is variable in time? This subject will be considered in Chapter 8 on predator–prey processes.]

Now notice that (3.13) is exactly of the same form as the original Verhulst differential equation (3.1a) such that term-by-term comparisons of the two equations reveal the equivalence:

$$\alpha \equiv \ln(r_m) \text{ and } \beta \equiv s(1-k).$$

But, what about the Pearl–Reed equation (3.1b) and its common version (3.1c)? Can we similarly interpret their ecological significance in comparison with (3.13)? Not quite. The term-by-term comparison of (3.1c) with (3.13) reveals the following:

$$\rho \equiv \ln(r_m) \quad (3.14a)$$

$$\rho/K \equiv s(1-k), \quad (3.14b)$$

and hence:

$$K \equiv \ln(r_m)/s(1-k). \quad (3.14c)$$

[Note: Don't forget that the parameter K is not the same as k, the latter being a measure of intensity in competition.] We see that the equivalence of (3.1c)–(3.13) is not as straightforward as the equivalence of (3.1a) to (3.13). In fact, the relationships in (3.14a, 3.14b, and 3.14c) reveal a serious problem hidden in the common notions (interpretations) of the parameters ρ and K, especially in the notion that K represents the carrying capacity of the environment. These need to be carefully reinterpreted.

3.7 Reinterpretations of Parameters ρ and K

3.7.1 Parameter ρ

As (3.14a) shows, the parameter ρ, commonly known as the 'intrinsic rate of increase', is related to the potential (maximum) reproductive rate r_m of an individual when it has no competitors. This makes good sense ecologically. But, what is the significance of ρ being mathematically equivalent to $\ln(r_m)$?

Consider what happens if $0 < r_m < 1$. We see that the instantaneous rate of increase (dx/dt) in (3.13) becomes negative because $\ln(r_m) < 0$, while $s(1 - k)x_t > 0$ as defined. This means that the population decreases consistently to extinction. Nothing is wrong about population extinction, as it happens all the time in the real world. Thus, a negative ρ in (3.14a) at first appears to indicate population extinction, as does (3.13). But, in fact, this is not always so in (3.1c).

Notice that a negative $\rho < 0$ implies $dx/dt < 0$ provided that the expression $(1 - x/K)$ in (3.1c) is positive, i.e. if the population x stays below K. But, what happens if x exceeds K? The expression $(1 - x/K)$ becomes negative. Then, for $\rho < 0$, the rate of increase (dx/dt) becomes positive, indicating a population increase. How can the population increase when its intrinsic rate of increase is negative? You might argue that, as K is the carrying capacity of the environment, x cannot exceed K by definition: that is, you might argue that this situation is improbable and need not be considered.

But is it really true? After all, the notion of K as 'the carrying capacity' was intuitively conceived to make sense of equation (3.1c) after it was given. In fact, there is no logical necessity that K is the limited capacity of the environment to support the population. So, let us critically examine the nature of the parameter K. For this purpose, I suggest that we first consider a situation in which ρ remains positive as usual, and then consider the situation in which ρ is negative.

3.7.2 Parameter K for $\rho > 0$

The equivalence (3.14c) exposes a problem in the conventional notion that K is the 'carrying capacity of the environment'. This is because none of the three parameters on the right-hand side of (3.14c), i.e. r_m (potential reproductive rate), s (individual resource requirement), and k (intensity of competition), is immediately pertinent to the environmental capacity. So, what is it? As some mathematicians and ecologists are aware, although not well publicized, K is in fact an equilibrium level of the logistic population process (3.1c). This can be readily verified.

Let x^* be the equilibrium density, and notice that, at equilibrium, there would be no change in the population. This means that, for $x = x^*$, the instantaneous rate of change $dx/dt = 0$, and substituting x^* for x in (3.1c), we find $K = x^*$.

Now that K is an equilibrium level, x can be larger than K. Let us see what happens. If x is so placed, i.e. $x/K > 1$, the instantaneous rate (dx/dt) in (3.1c) becomes negative for $\rho > 0$, which means a decrease in population. More precisely, the population decreases asymptotically (and monotonically) to the level K: how fast or slowly it decreases depends on the value of $\rho > 0$; the lower the value the slower, of course. [Suggestion: I encourage you to draw a graph, using equation (3.3c), to confirm the above tendency.] It implies that, once placed above K, the population x stays above K all the time, although decreasing asymptotically to K: it may decrease at an arbitrarily slow rate if ρ is so chosen. How come the population can stay for so long beyond the level where the environment is supposed to be unable to support it?

This demonstrates that K is not the maximum level of the population that the environment can support, but it is (on the condition $\rho > 0$) the stable equilibrium level of the common-version logistic process: whenever displaced to either side of K, the population returns asymptotically to it. [Note: In deriving the differential equation (3.1c), it was assumed that $0 < x < K$, and you might have thought that K meant to be the uppermost limit. This is not necessarily so, in that K acts as *an upper limit* whenever x grows from below K, inasmuch as K would act as *a lower limit* of x whenever x starts from above K, whereas the lowest (realizable) limit of x is 0.]

After all, except for the semantic problem in the ecological meaning of K, the usual (and well-known) logistic law would follow the model (3.1c), provided that ρ is positive. I now examine the nature of K when ρ becomes negative.

3.7.3 Parameter K for ρ < 0

Consider that the population x is placed exactly at the equilibrium level K. Then, the rate of change (dx/dt) of model (3.1c) is equal to zero, meaning no change in x: it stays at K for good if undisturbed. Now, consider that x is slightly less than K, so that $(1 - x/K)$ is positive. But, multiplied by a negative ρ, the rate of change (dx/dt) becomes negative, meaning a population decrease; it decreases consistently to extinction. Nothing is wrong with population extinction; it always happens. However, a problem arises if x is placed slightly above K, as shown below.

Because $x > K$ means $(1 - x/K) < 0$, when multiplied by a negative ρ, the rate of increase $(dx/dt)/x = \rho(1 - x/K)$ becomes positive so that x increases away from K. In fact, the population in (3.3c) increases without bound even faster than the simple Malthusian geometric progression. You can readily confirm this by simulation with $\rho < 0$ and $x_o > K$. Try it, and you would find the process (3.3c) generates a weird pattern of dynamics: x increases first like a Malthusian process, but at a certain point in time, suddenly plunges to way below the level 0, i.e. x becomes negative; but then it starts to increase again and tends asymptotically to 0. Appendix 3D explains how this happens. Thus, altogether, we see that the population diverges from K whenever displaced from it on either side. In other words, under the condition that ρ is negative, the equilibrium K becomes unstable and the resultant pattern of dynamics would become unrealistic.

In contrast, the Verhulst process (3.3a) — as equivalent to the theoretical model (3.13) — would not exhibit this anomalous behaviour for $a = \ln(r_m) < 0$: no matter where it is placed initially, the population x simply decreases to 0 asymptotically, as a normal process of extinction. Appendix 3E shows the mathematics behind this attribute of (3.3a).

But why does model (3.3c) differ from (3.3a) in its response to changes in the sign (+ or −) of the parameter $\rho \equiv \ln(r_m)$? The answer lies in the structural problem in (3.1c).

3.8 Structual Problem of the Common-Version Logistic Model

Even though closely similar to the theoretical version (3.13), the common version (3.1c) differs in one aspect: in (3.13) there are three parameters in two independent groups, i.e. $\ln(r_m)$ and $s(1 - k)$, whereas in

(3.1c) there are two parameters in two numerically linked groups, i.e. ρ and ρ/K. This structural attribute of (3.1c) creates the problem: when ρ is changed from positive to negative, it automatically changes the expression ρ/K to negative, whereas in (3.13), the value of r_m can be changed without affecting either the value or the ecological meaning of the expression $s(1 - k)$. In other words, the common version is unable to handle the situation in which $0 < r_m < 1$, and this limits the practical applications of the model. In particular, it is unsuitable for the analysis of an endangered species in which the potential reproductive rate r_m might become less than 1 under some external environmental influences, as will be discussed in Chapter 4. Altogether, in view of the fact that the common version of the logistic model is no simpler in structure than the theoretical version, there is no advantage of using it for practical purposes. So, I suggest that this epoch-making work of Pearl and Reed be retired in a history museum.

Now, to finalize the present chapter, I should remark on a few more issues to be kept in mind when studying the logistic law.

3.9 Final Remarks of Chapter

First of all, I never imply that any of the past works that have been based on the common-version logistic equation (either 3.1c or 3.3c) be reconsidered. For instance, I see no problem with those works in which the common version (3.3c) was successfully fitted to the observed process, i.e. no problem with those works in which the estimated value of ρ were positive. As a matter of fact, I have never seen a work in which ρ was estimated to be negative. This is presumably because the logistic model (of any version) has always been fitted to increasing populations from below K: I imagine that, quite naturally, few people have attempted to fit the model to a population that follows the Malthusian progression or is on the verge of extinction.

Second, it must be true that an environment has the maximum capacity to support a population: it is even axiomatic in that it never happens that this planet becomes full of elephants or fleas. The problem is, however, that we would not know where the maximum limit lies: we would not even know how to define, let alone quantify, what the capacity is from a pragmatic point of view. Notwithstanding, I am inclined to think it is unlikely that populations of many animals would stay even close to resource limits. Thus, I insist that those works (arguments) which were based on the assumption of K being the uppermost

52 · Logistic Law of Population Growth

limit should be reconsidered. Also, those arguments which are based on the assumption that ρ and K are independent parameters should be critically re-examined.

Third, for a descriptive device, I insist on the form the theoretical version (3.13) takes, the original Verhulst version being an example. However, for practical applications, I would use the discrete-time version (3.10) because, as will be demonstrated in Chapter 4, it generates a variety of dynamical (time-dependent) patterns, whereas (3.13) as a continuous-time model generates only one pattern: asymptotic conversion to an equilibrium level.

Finally, beyond descriptive purposes, the theoretical model (3.10) serves as an analytical tool and provides a basis for constructing models that reveal the ecological mechanisms underlying population processes on the higher levels of complexity. I deal with these issues in subsequent chapters, one at a time.

Appendix 3A: How to Solve a Differential Equation in the Models (3.1)

Consider the (unknown) function $f(u)$ of a generic variable u and its derivative $df(u)/du$, concisely written as $f'(u)$, i.e. $df(u)/du \equiv f'(u)$. [Recall that the symbol '$\equiv$' reads 'identical or equivalent to'.] This can be rearranged to the form:

$$df(u) = f'(u)du. \tag{3A.1}$$

We are interested in finding $f(u)$, given $f'(u)$ as a known function. This operation is called 'integration', a reversal of differentiation which is to find $f'(u)$, given $f(u)$. In other words, each side of (3A.1) can be integrated to find $f(u)$. An integral is usually written using the symbol $\int$ (an elongated letter S for summation) such that its application to (3A.1) gives:

$$\int df(u) = \int f'(u)du = f(u) + c \tag{3A.2}$$

where c is the constant of integration, the meaning of which will become apparent shortly.

To apply formula (3A.2) to the (common-version) differential equation (3.1c) in the main text, we first rearrange (3.1c), on the condition that $1 - x/K \neq 0$, into the form:

$$[1/(1 - x/K)x]dx = \rho dt. \tag{3A.3}$$

Further, with a little algebraic manipulation, we can modify the expression $1/(1 - x/K)x$ on the left-hand side as the sum of two parts, i.e. $1/(1 - x/K)x = [1/(K - x)] + (1/x)$, such that:

$$[1/(1 - x/K)x]dx = [1/(K - x)]dx + (1/x)dx.$$

Hence, we can integrate each side of (3A.3) to find:

$$\int [1/(1 - x/K)x]dx = \int [1/(K - x)]dx + \int (1/x)dx = \rho \int dt. \quad (3A.4)$$

Generally speaking, the process of integration is technically quite cumbersome. Fortunately, the integrals of many (standard) functions have been worked out in mathematics, and are listed in many tables which are readily available. So, you can pick up one of them to find:

$$\int dt = t + c_1 \quad (3A.5i)$$

$$\int (1/x)dx = \ln(x) + c_2, \; x > 0 \quad \text{[Note : 'ln' stands for 'natural logarithm'.]} \quad (3A.5ii)$$

$$\int [1/(K - x)]dx = -\ln|K - x| + c_3, \; 0 < x \neq K \quad (3A.5iii)$$

where c_1, c_2 and c_3 are the constants of integration which I shall explain in a moment. Also, the expression $|K - x|$ in (3A.5iii) is an absolute value of the difference $(K - x)$. This means that we take whichever value of $(K - x)$ is positive. For instance, if $(K - x) > 0$, the solution on the left-hand side of (3A.5iii) is equal to $-\ln(K - x)$, ignoring the constant c_3. If $(K - x) < 0$, then $-(x - K) > 0$ such that equation (3A.5iii) can be written: $-\int [1/(x - K)]dx = -\ln(x - K)$. Thus, the right-hand side of (3A.5iii) is a compact way of writing whichever is positive. In formulating the differential equation (3.1c) in the main text, we assumed that $0 < x < K$, or $(K - x) > 0$. Therefore, we need not be bothered by the absolute sign, at least for the moment; see [Notes] after (3A.7) below.

Thus, substituting the right-hand sides of (3A.5, i, ii, iii) for the corresponding terms in (3A.4), gathering c_1, c_2 and c_3 on the right-hand side (writing it as c), and using the rule that $\ln(m) - \ln(n) = \ln(m/n)$, we find a general solution of the differential equation (3.1c) in the main text to be:

$$\ln[x/(K - x)] = \rho t + c. \quad (3A.6)$$

54 · Logistic Law of Population Growth

Now, let us evaluate the constant c. To do this, assume that we start the population at an arbitrary size $x_o > 0$, i.e. $x = x_o$ when $t = 0$. Substituting these in (3A.6), we find:

$$\ln[x_o/(K - x_o)] = c.$$

Further, substituting $\ln[x_o/(K - x_o)]$ for c in (3A.6), relocating it to the left-hand side, and using the rule $\ln(m) - \ln(n) = \ln(m/n)$ again, we have the complete solution with the arbitrary initial state x_o:

$$\ln\{[x/(K - x)]/[x_o/(K - x_o)]\} = \rho t.$$

After an anti-ln transformation of the above relationship, we have:

$$[x/(K - x)]/[x_o/(K - x_o)] = e^{\rho t}. \tag{3A.7}$$

Solving (3A.7) for x yields the (common-version) logistic equation (3.3c) in the main text.

[Notes: Although having started with the assumption $0 < x < K$, I shall, in the later section of the present chapter, consider a situation in which $x > K$, which means that, mathematically, x_o can be larger than K. However, as will be shown, x always stays on the same side of K where it started as x_o. So, the difference $(K - x)$ always has the same sign as $(K - x_o)$. Thus, the quotient on the left-hand side of (3A.7) would stay positive. Therefore, the logistic equation (3.3c) in the main text holds even if $x_o > K$.]

But what if $x = K$? This violates the condition on which (3A.3) was formulated: the integral (3A.5iii) would not exist, and it appears that the formula (3A.6) cannot be a solution. Does this mean that there is no solution for the differential equation (3.2) for $x = K$? Yes, there is. Let us go back to (3.1c) in the main text to find that, for $x = K$, the rate of change $(dx/dt) = 0$. This means that, once x has reached K (which never happens within a finite period of time, but you may place x_o at K), there would be no further changes in population: x would thenceforth stay there at K. In other words, $x = K$ is the solution of (3.1c) for $x = K$: the condition itself is the solution, as it were. Thus, the logistic equation (3.3c) holds for all $x > 0$. [Note: In ecology, we need not consider what happens if $x = 0$ because nothing would happen to nothing.]

The original Verhulst differential equation (3.1a) can likewise be solved to obtain (3.3a) in the main text: technically, just replace ρ with α and K with α/β.

Appendix 3B: The Derivative $d(e^u)/du = e^u$

Let me first introduce the formal definition of the derivative of a generic function, say $f(u)$, of the generic variable u. Consider that a given section of the curve formed by plotting $f(u)$ against u is smoothly continuous; that is, there is neither a break nor a sharp bend within the section. Consider now a small increment in u, say Δu, starting at a given point of u within that section of the curve. Then, there is a correspondingly small increment in the function from $f(u)$ to $f(u + \Delta u)$, written $\Delta f(u)$. That is, $\Delta f(u) = f(u + \Delta u) - f(u)$. Now, let the increment Δu become infinitesimal, and accordingly becomes $\Delta f(u)$ infinitesimal. Then, the symbol Δ is replaced by the letter d, i.e. du and $df(x)$. These are called 'differentials' and the expression $df(u)/du$ is called the 'derivative' of the function $f(u)$ with respect to u. So, we have the definition of the derivative:

$$df(u)/du = [f(u + \Delta u) - f(u)]/\Delta u \text{ in the limit } (\Delta u \to 0). \quad (3B.1)$$

In the present case, $f(u)$ is e^u. Then, the expression on the right-hand side of (3B.1) is equal to $(e^{u+\Delta u} - e^u)/\Delta u$. Thus,

$$d(e^u)/du = (e^{u+\Delta u} - e^u)/\Delta u \text{ in the limit } (\Delta u \to 0). \quad (3B.2)$$

But, by the rule $e^{n+m} = e^n e^m$, we can rewrite the right-hand side of (3B.2) as:

$$e^u(e^{\Delta u} - 1)/\Delta u. \quad (3B.3)$$

Now, as Δu approaches ever more closely to 0, and because $e^0 = 1$, both numerator and denominator in (3B.3) approach 0 ever more closely and, as you know, the ratio 0/0 is indeterminate. We somehow need to evaluate this limiting form. There are several different ways, but here I use the identity discovered by the seventeenth-century Swiss mathematician Jacob Bernoulli:

$$e = (1 + \Delta u)^{(1/\Delta u)} \text{ in the limit } (\Delta u \to 0).\text{[See Note below.]} \quad (3B.4)$$

Then, $e^{\Delta u} = [(1 + \Delta u)^{(1/\Delta u)}]^{\Delta u} = (1 + \Delta u)$, and hence:

$$(e^{\Delta u} - 1)/\Delta u = (1 + \Delta u - 1)/\Delta u = \Delta u/\Delta u = 1.$$

Substituting this result in expression (3B.3), we find its limiting form to be simply e^u. Thus, the derivative $d(e^u)/du$, defined by (3B.2), is equal to e^u, i.e. the derivative of e^u is itself.

[Note: By the way, you may alternatively write $1/\Delta u = n$ in (3B.4) such that e can also be defined as: $e = (1 + 1/n)^n$ in the limit $(n \to \infty)$. You can use whichever expression is convenient.]

Apparently, Bernoulli found that the expression (in either form) would converge to a certain constant in the limit ($\Delta u \to 0$ or $n \to \infty$), to which the letter e was later assigned by another Swiss mathematician, Leonhard Euler. Thus, the number e is known as 'the Euler number' among mathematicians. But the number e is most popularly known among ecologists as the base of the natural logarithm, its origin being attributed to John Napier of sixteenth-century Scotland.]

Appendix 3C: The Derivative $d(\ln v)/dv \equiv 1/v$, $v > 0$

Let $\ln(v) = u$ which, by inversion, is equivalent to $v = e^u$. Then, $dv/du = d(e^u)/du = e^u$ by the result in Appendix 3B. Then, $d(\ln v)/dv = du/dv = 1/(dv/du) = 1/e^u = 1/v$. Voila!

Appendix 3D: The Anomaly in the Common Version Logistic Equation (3.3c)

Equation (3.3c) in the main text is given in the form: $x = x_o e^{\rho t}/[1 + (e^{\rho t} - 1)x_o/K]$. For convenience, I modify it by dividing the denominator and numerator by $e^{\rho t}$ such that, after a little algebraic manipulation, we have:

$$x = Kx_o/[e^{-\rho t}(K - x_o) + x_o]. \qquad (3D.1)$$

Let us see what happens to the denominator (D, say) of (3D.1) for $\rho < 0$ (hence $-\rho t > 0$) and $0 < K < x_o$. At $t = 0$, $e^{-\rho t} = 1$, and hence $D = K > 0$. As t increases, $e^{-\rho t} > 1$ would, after a while, become so large that D becomes negative because $(K - x_o) < 0$. Therefore, there must be a point in t, say t_s, at which $D = 0$. We see that $t = t_s$ is a singular point at which x is discontinuous. In particular, if t approaches t_s from below, x tends to ∞. But, as soon as t passes t_s, x becomes negative infinity, although from then on continuously increases as t increases further, tending asymptotically to 0 from below.

Appendix 3E: Mathematical Attributes of the Verhulst Equation (3.3a)

The denominator, $[1 + (e^{\alpha t} - 1)(\beta/\alpha)x_o]$, of equation (3.3a) in the main text stays positive regardless of α being positive, 0, or negative. Obviously, for a positive α, the denominator is positive. Even for $\alpha < 0$, the quotient

$(e^{\alpha t} - 1)/\alpha$ in the denominator stays positive because the expression $(e^{\alpha t} - 1)$ is also negative. For $\alpha = 0$, the quotient becomes $0/0$, i.e. indeterminate. However, this can be evaluated as a limiting case: that is, in the limit α tends to 0 (regardless of α tending to 0 from the negative or positive side), the quotient converges consistently to t. [Note: You can confirm this attribute by the l'Hôpital rule explained in Chapter 4.] Thus, for $\alpha \leq 0$, the expression $e^{\alpha t}$ in the numerator of (3.3a) converges to 0 as t increases and, accordingly, the population x becomes extinct as happens in the real world.

4 · *Reproduction Curves and Their Utilities*

4.1 Original Ideas

The late Professor Patrick A. P. Moran of Australian National University was a statistician who was deeply interested in biology and made particularly important contributions to population ecology and genetics. While he was a research fellow in statistics, Pat published a paper entitled 'Some remarks on population dynamics' (Moran, 1950). To illustrate one of several points in the article, Pat used a curve, a plot of population size in one generation against that of the previous generation in a single-species population without age-class distinctions. He did not use a particular mathematical model but drew an idealized curve by hand for illustration. The graph did not show axis designations, no tick marks, no legend, but just said 'figure 1' at the bottom. Explanations were all incorporated in the text. I reproduce his graph here in Figure 4.1a: not an exact but a freehand copy. However, I have added axis names to suit me; you only need to assume that the unit of the scale (population size x) is the same on both axes, such that $x_{t+1} = x_t$ on the 45° diagonal line.

The utility of this graph is the following. Starting with a founding population, say x_1, on the (horizontal) x_t-axis in Figure 4.1b, draw an upward vertical line on x_1 until it intersects the curve. From the intersection, draw a horizontal line to the left until it intersects the vertical (x_{t+1}) axis. The intersection reads the size of the following generation, x_2. Now, extend the horizontal line to the right until it intersects the diagonal (45°) line. From the intersection, draw a vertical line downwards. Its intersection with the x_t-axis gives x_2: that is, x_2 on the x_{t+1}-axis is transferred to the x_t axis. By repeating the same procedure but starting from x_2 on the x_t-axis, you find x_3 on the x_{t+1}-axis, transfer it onto the x_t-axis, and so on and on to recursively generate a series of population changes over many generations. By the way, at the point where the curve intersects the 45° diagonal line, the population at any generation stays unchanged because, at the point, $x_{t+1} = x_t = x^*$, say. Thus, the

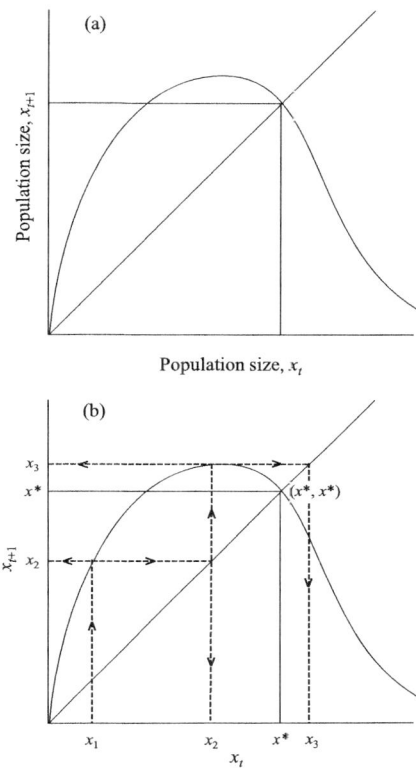

Figure 4.1 (a) An idealized reproduction curve as originally envisaged by Moran (1950). (b) How to graphically generate a population series $\{x_t\}$ (with permission from John Wiley & Sons/Blackwell Publishing Ltd; International Biometric Society).

coordinates of the intersection, $(x_t = x^*, x_{t+1} = x^*)$, or simply (x^*, x^*), marks an equilibrium point. [Professor Mark Williamson of the University of York, with whom I have shared much interest in many aspects of population biology, called the Figure 4.1 curve the 'Moran plot' in his book (Williamson, 1970).]

Shortly after Moran's 1950 paper, the Canadian fisheries biologist William E. Ricker published his magnum opus 'Stock and recruitment' (Ricker, 1954). In this article, Ricker studied relationships between the stock (parental) and recruited (offspring) populations. He used many graphs in which the size of the recruited is plotted against the stock, the idea essentially the same as the Moran plot. [I do not know if Ricker was aware of Moran's paper; he did not quote it.] Ricker, too, did not use a particular mathematical model but varieties of hand-drawn curves, using his imagination; some of them looked much the same in shape as the Moran plot. Ricker aptly called those curves the 'reproduction curves'.

Despite an apparent similarity, however, there is a significant difference between Moran and Ricker in their uses of the curves. Basically, the Moran plot directly applies to the process of a population with discrete generations, each comprising a single age-class like many insects in which every individual dies after eggs have been laid. Therefore, the size of a given generation depends only on that of the immediately preceding generation. In other words, we can generate a series of population sizes, generation after generation, by recursively reading them on the curve. Thus, the Moran plot is a self-contained system.

On the other hand, Ricker was dealing with fish populations in which generations overlap: because fish are in general long-lived, the stock population of a given year is made up of individuals of different ages: evidently, this also applies to bird and mammal populations. Thus, the recruited population in a given year that is assessed by a reproduction curve comprises only the youngest age-class of the stock population, which would require several years to reach reproductive maturity. Therefore, in order to generate a series of the population as a whole, the recruited generation of a given year must somehow be incorporated into the stock population. In addition, variations in the annual reproductive rate, and mortality among different age-classes of the stock population, must also be considered. In other words, Ricker's reproduction curve is not a self-contained system.

To deal with this problem, Ricker constructed a supplementary (numerical) table of idealized age structure to form the stock population for the following year, an idea similar to the Leslie matrix (Leslie, 1948, not cited in the Ricker article). With this supplementary table, Ricker assessed the size of the new stock population for the following year to be incorporated into the reproduction curve and continued to graphically generate the series of annual population changes.

However, the age-class structure depends on individual cases and taking it into account could undesirably complicate the way we find the utility of a reproduction curve beyond the level I intend to consider here. So, in the present book, I adopt the term 'reproduction curve' after Ricker but use it in the sense of the Moran plot for a self-contained process of a single-species population. Thus, the population x_t may be comprised of different age-classes, while x_{t+1} is in general made up of the survivors of x_t and their newly recruited (born) offspring.

4.2 Drawing a Reproduction Curve

Let us draw a reproduction curve, using the model (3.10) developed in Chapter 3. To repeat it:

$$r_t = r_m \exp[-s(1-k)x_t] \quad (3.10 \text{ rpt})$$

where $r_t = x_{t+1}/x_t$ is the (mean) per-capita (net) rate of change in population; r_m is the (mean) potential reproductive rate of an individual (when it has no competitors); and $s(1-k)$ is a measure of the intensity of competition (given the resource requirement s) among the individuals of the population. [In the following, the adjective (mean) will be dropped.] Thus, writing $s(1-k)$ compactly as c, we have:

$$r_t = r_m \exp(-cx_t) \text{ or } x_{t+1} = x_t r_m \exp(-cx_t). \quad (4.1)$$

Using the compact model (4.1), let us draw a reproduction curve with the parameter values ($r_m = 2.5$, $c = 1.0$) chosen conveniently for the graphics. With a computer software package that does some simple calculations (e.g. Minitab), try the following: store the constants $r_m = 2.5$ and $c = 1.0$; set a series of numbers as x_t in one column, ranging from 0 to 5 at a 0.01 interval, say; and let the computer calculate x_{t+1}; the 0.01 interval makes the resultant curve quite smooth. Then, plot x_{t+1} on the vertical axis against x_t on the horizontal axis to form the reproduction curve (Moran plot) of the model (4.1): make sure that the unit of the two axes is physically equal in length. Designate this graph Figure 4.2a. In addition, draw a 45° diagonal line through the origin (0, 0) until it crosses the curve. You see the 45° line intersect the curve at $x_t = x_{t+1}$, which is the equilibrium point (x^*, x^*) as in Figure 4.1b.

We are now ready to generate the series of x_t over $t = 1, 2, 3, \ldots$, written $\{x_t; t = 1, 2, 3, \ldots\}$ or just the series $\{x_t\}$, in the manner illustrated in Figure 4.1b. For convenience, the section of the graph in Figure 4.2a (ranging from 0 to 1.0 on both axes) is enlarged in Figure 4.2b.

4.3 Generating the Population Series

Starting with x_1 arbitrarily placed on the x_t-axis close to the origin (0, 0) of the coordinate system, the curve in Figure 4.2b recursively generates the series $\{x_t\}$ like ascending steps (solid lines with arrows) between the 45° line and the reproduction curve. By the time the series has come very close to the point (x^*, x^*) where the reproduction curve intersects the 45° line, the steps are squeezed so much into the narrow space between

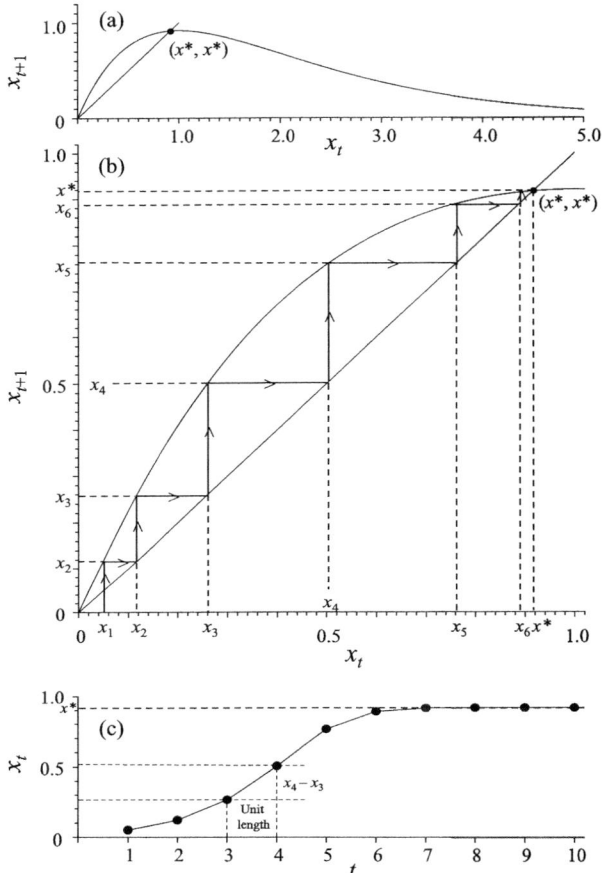

Figure 4.2 (a) A reproduction curve that model (4.1) generates with the parameter values $(r_m, c) = (2.5, 1.0)$. (b) Graphical method of generating a population series in the manner illustrated in Figure 4.1b. (c) The population series thus generated.

the curve and the line that further reading becomes difficult. Nonetheless, you can readily visualize that the series $\{x_t\}$ would get closer and closer to eventually converge to x^*. In other words, with the parameter values $(r_m = 0.25, c = 1.0)$ and starting off with x_1 below x^*, the series $\{x_t\}$ converges (in the limit) to x^* and would stay there forever if undisturbed. That is, x^* is an equilibrium density which is approximately equal to 0.9163. [Note: How do I know this? Notice that once the equilibrium has been reached, there will be no change in population, which means $r_t = 1$ regardless of t. So, substituting 1 for r_t

and x^* for x_t in model (4.1), you have $1 = r_m\exp(-cx^*)$, or $\exp(cx^*) = r_m$, and solving the result for x^*, we find $x^* = [\ln(r_m)]/c$. However, in the present example, $r_m = 2.5$ and $c = 1.0$. Hence, $x^* = \ln(2.5) \approx 0.9163$.]

Plotting the series $\{x_t\}$ against t in Figure 4.2c, we see the familiar sigmoid growth curve, just like the one that the classical logistic model exhibits. No surprise! As already shown in Chapter 3, the classical model is a continuous-time version of the discrete-time model (3.10) which is compactly written and re-coded as (4.1) in the present chapter. You can see extra remarks in Figure 4.2c between the third and fourth time steps. I will get to them shortly.

It should be noted that the series in Figure 4.2c is just one example that the model (4.1) generates with the particular set of parameter values (r_m, c) = (2.5, 1.0). [Note: This expression of equality is a compact way of writing: $r_m = 2.5$ and $c = 1.0$.] You would see a variety of patterns of population series by changing the parameter values of the model: these will be shown in a later section in the current chapter.

Now, the variation in population series depends totally on the variation in the shape of the reproduction curve. So, it would be a good idea to be acquainted with the mathematical roles that the parameters (r_m, c) play in controlling the shape of the reproduction curve. It might be rather tedious work but it is essential in doing quantitative ecology: once used to it, you would be all set. So, here we go!

4.4 Mathematical Roles that the Model Parameters Play

Recall that the ecological meaning of the parameter r_m is the potential reproductive rate of an individual when it has no competitors, and $c \equiv s(1 - k)$ is a measure of the intensity of (intraspecific) competition, given the resource requirement s. [The symbol '$\equiv$' reads as 'identical to'.] But what do these parameters do mathematically? They determine the shape of the reproduction curve. For instance, r_m determines the slope of the curve when starting at the origin (0, 0), whereas c is concerned with determining the position of the apex, and so on. We can find those attributes by tracing the curve analytically as shown below. Let us begin with some basics.

4.4.1 Basics

Consider a smoothly continuous curve, i.e. the plot of $f(x_t)$ vs. x_t without breaks or sharp bends, like the one in Figure 4.2a: see [Note] below. We can trace the shape of the curve by looking into its slope at a given point. But what is the slope of the curve? It is the slope of the tangent line at the given point. [Note: The tangent line is the line that just touches the bulging side of (but without crossing) the curve at or in the close vicinity of the given point.] Thus, the slope of the curve is measured by the slope of the tangent line, given by the trigonometric function tan θ in which θ is the angle that the tangent line makes with the (horizontal) x_t-axis. [Note: Although x_t is the size of population measured at a discrete interval of time, its value varies continuously from 0 to infinity such that the curve Figure 4.2a, i.e. x_{t+1} plotted against x_t, is continuous without sharp bends or breaks.]

Now, let us select a point arbitrarily at $x_t = a$, say. Then, tan θ is evaluated by the derivative $df(x_t)/dx_t$ of the curve at the point $x_t = a$, which is compactly written as $f'(x_t = a)$ or just $f'(a)$. In short, tan $\theta = df(x_t)/dx_t = f'(x_t)$. In particular, a positive $f'(a)$ means a positive slope, i.e. the curve is increasing; a negative $f'(a)$ indicates a negative slope, i.e. the curve is decreasing; and, in between, $f'(a) = 0$ indicates that the slope is flat, e.g. at the apex (peak) or trough of the curve. [Note: Appendix 4A gives graphical explanations of the above attributes. So, if you are not familiar with derivatives in calculus, or have forgotten about them, I suggest that you quickly go through the Appendix to understand what these attributes are, rather than just swallow them as given.]

4.4.2 Tracing the Figure 4.2a Curve as an Example

Now that the equation of the curve is given by (4.1), i.e. $x_{t+1} = x_t r_m \exp(-cx_t)$, we want to find its derivative to trace the curve. To this end, let us write the equation compactly as $x_{t+1} = f(x_t)$ such that $dx_{t+1}/dx_t = df(x_t)/dx_x = f'(x_t)$. Then, look into Appendix 4B which shows how to derive $f'(x_t)$ for (4.1). The result is given by:

$$f'(x_t) = r_m(1 - cx_t)\exp(-cx_t) \\ = r_m(1 - cx_t)/\exp(cx_t), \quad \because \exp(-cx_t) \equiv 1/\exp(cx_t). \quad (4.2)$$

[Notes: The symbol "$\because$" reads 'because'. The symbol '$\equiv$' reads 'identical' or 'equivalent'.] We are now ready to trace the curve by looking into how $f'(x_t)$ in (4.2) changes as x_t increases from its origin.

Start at the origin. By definition, $x_t = 0$ at the origin. Substituting $x_t = 0$ on the right-hand side of (4.2), and noting that $\exp(0) \equiv e^0 = 1$, we find $f'(0) = r_m$, i.e. tan $\theta = r_m > 0$ at the origin of the curve. That is, the curve starts rising at the angle $\theta = \tan^{-1}(r_m)$. [Notes: The notation '$\tan^{-1}(r_m)$' reads 'arctangent of r_m', the inverse of $\tan \theta = r_m$. To evaluate the angle θ, you need not calculate it from scratch: just use your pocket calculator. For example, with the curve in Figure 4.2a in which $r_m = 2.5$, your calculator tells: $\tan^{-1}(2.5) \approx 68.2°$ or $\approx 0.379\pi \approx 1.19$ in radians.]

Position of the apex. Starting at the origin, the curve increases until it reaches its apex at which its slope becomes flat: that is, $\tan \theta = f'(x_t) = 0$ in (4.2), i.e. $(1 - cx_t) = 0$ or $x_t = 1/c$. Thus, the curve reaches its apex at the point $x_t = 1/c$. But how high is the apex? Obviously, the height is x_{t+1} at $x_t = 1/c$. So, substituting $1/c$ for x_t in equation (4.1), you find: the height $= r_m/ce$ in which $e = 2.71828\ldots$ is the base of natural logarithms (or the Euler number). However, in the Figure 4.2a curve, $r_m = 2.5$ and $c = 1$. Hence, the height is $2.5/e \approx 0.92$.

After the apex. Having passed its apex, i.e. $x_t > 1/c$ or $cx_t > 1$, where would the curve go? Notice first that $(1 - cx_t) < 0$ there, whereas $r_m/\exp(cx_t) > 0$, and hence, from (4.2), $f'(x_t) < 0$ everywhere, i.e. the curve is decreasing everywhere. But where does it decrease to eventually? Take a look again at the right-hand side of (4.2), i.e. $r_m(1 - cx_t)/\exp(cx_t)$. As x_t increases further and further, the denominator $\exp(cx_t)$ tends to positive infinity, whereas the numerator $(1 - cx_t)$ tends to negative infinity. Thus, the quotient becomes $-\infty/\infty$, i.e. indeterminate. Wow! What shall we do? Fortunately, we have a trick known as l'Hôpital's rule, named after the seventeenth-century French mathematician Guillaume François Antoine, Marquis de l'Hôpital; quite a mouthful. The trick is the following.

Consider the two functions $h(u)$ and $g(u)$ of a generic variable u. Consider also that both functions are 'well-behaved' (a mathematical jargon for 'differentiable', meaning smoothly continuous) at (and in the vicinity of) a given value, $u = a$, say. Consider further that each function tends to positive or negative infinity in the limit $u \to a$. Then, l'Hôpital's rule states that: the quotient $h(u)/g(u)$ is equal to $h'(u)/g'(u)$ in the limit $u \to a$; Appendix 4C gives details. So, if at least one of these derivatives turns out to be finite-valued (constant) at $u = a$, then we can evaluate the quotient. In the present case, writing $(1 - cx_t) = h(x_t)$ and $\exp(cx_t) = g(x_t)$, and following the procedures in Appendix 4B, we find: $h'(x_t) = -c$ and $g'(x_t) = c\exp(cx_t)$. Thus, we find that, in the limit $x_t \to \infty$, $h'(x_t)/g'(x_t) = -1/\infty = 0$. In other words, the decreasing curve becomes flatter and flatter as its skirt extends to the right indefinitely and eventually converges asymptotically to the 0 level (i.e. the horizontal x_t-axis). We now go back to the Figure 4.2a curve.

Inflection point. After having passed the apex, the curve continues to decrease but in the following manner: first, it decreases at an increasing (faster) rate until a certain point and, thenceforth, decreases at a slower rate before converging to the x_t-axis. So, we see that there must be an inflection point somewhere. We can determine the point, using the method described in the last paragraph in Appendix 4A: the second derivative $f''(x_t) = 0$ at an inflection point. So, the evaluation of $f''(x_t)$ is in order. We have already evaluated the first derivative $f'(x_t)$ in (4.2). Thus, consulting Appendix 4B and differentiating $f'(x_t)$ for the second time, we find:

$$f''(x_t) = cr_m(cx_t - 2)/\exp(cx_t). \qquad (4.3)$$

Equating the right-hand side to 0, we find $(cx_t - 2) = 0$, i.e. $x_t = 2/c$. But because $c = 1$ as assumed in the present example, we find the inflection point to be at $x_t = 2$. This completes the analyses of the mathematical roles that the parameters play in determining the reproduction curve that the model (4.1) generates.

The foregoing representation of a reproduction curve in Figure 4.2a and b (i.e. the plot of x_{t+1} vs. x_t) is a natural way to introduce its utility: the graphical method of generating the population series. However, it is not of ideal form for practical purposes. For one thing, the process of generating the series $\{x_t\}$ graphically, as in Figure 4.2b, is rather cumbersome. For another, and more seriously, the time-plot of $\{x_t\}$, as in Figure 4.2c, may at times be visually deceptive (illusory). These problems arise primarily from the fact that the variable x, representing the size of a population, is a non-negative quantity, even though a negative person may exist in a human population.

4.5 Problems with Population Size as a Non-negative Quantity

Glancing at the familiar sigmoid (S-shaped) curve of the logistic model in many textbooks, you might have thought, or might possibly have been taught, that: starting with a low density, the population would grow slowly at first, then rapidly, and would again slow down to level off. Is this interpretation correct? Yes, it is correct mathematically. No, it is not appropriate from the ecological point of view.

Mathematically, we may measure the rate of change in population per unit time. In fact, the conventional textbook interpretation (perception) of the 'sigmoidality' is based tacitly on the *per-unit time* rate. [Note: I have seen

4.5 Population Size as a Non-negative Quantity

no word like 'sigmoidality' in standard dictionaries. I happened to come across an article in which its author used the word. I have no idea if he invented it. In any case, it conveys its meaning succinctly. So, why not use it?] In contrast, my approach is based on the *per-capita* rate because it is ecologically more meaningful: in fact, it is a fundamental mechanism underlying any population process. But, what difference would they actually make?

The per-unit time rate of change in population from generation t to $t+1$ is the difference between x_t and x_{t+1}, i.e. $(x_{t+1} - x_t)$ as we see in Figure 4.2c for $t = 3$ to 4. As opposed to this, the per-capita rate of change is measured by $r_t = x_{t+1}/x_t$ as in model (4.1). The table below gives a numerical example of the difference between the two distinct ways of expressing the rate of change, using the data of the sigmoid curve in Figure 4.2c; all figures are rounded to the nearest 1000th:

(a) t	1	2	3	4	5	6	7	8	9	10
(b) x_t		0.050	0.119	0.264	0.507	0.763	0.889	0.914	0.916	0.916	0.916
(c) x_{t+1}		0.119	0.264	0.507	0.763	0.889	0.914	0.916	0.916	0.916	0.916
(d) x_{t+1}/x_t		2.378	2.220	1.920	1.506	1.165	1.027	1.003	1.000	1.000	1.000
(e) $x_{t+1} - x_t$		0.069	0.145	0.243	0.256	0.126	0.024	0.002	0.000	0.000	0.000

We see that the *per-capita* rate of change (d) takes the highest value at $t = 1$ and monotonically diminishes to 1 as the series $\{x_t\}$ approaches ever more closely to the equilibrium level, $x^* \approx 0.916$. As compared with this, the *per-unit time* rate of change (e) starts with a low value at $t = 1$, increases to the highest value at $t = 4$, and then decreases practically to 0 by $t = 7$, the pattern corresponding to the sigmoid growth of the series $\{x_t\}$ as in (b): it starts increasing slowly; exhibits the steepest ascent (the highest rate of increase) in the midway point in time ($t = 4$); and slows down again for a further increase in time. We are now ready to look into the problem with this conventional perception of the sigmoid growth curve.

4.5.1 Problem with the Sigmoid Growth Curve

The foregoing example reveals that the sigmoid curve hides information on the per-capita rate of change (x_{t+1}/x_t) for a very low level of population. This is because, when the population is low, the per-unit time rate ($x_{t+1} - x_t$) is low regardless of the value of x_{t+1}/x_t. That is, we just cannot tell if the initial section of the sigmoid curve carries something ecologically significant or not: only when the population approaches the equilibrium level where $x_{t+1} = x_t = x^*$ does a low value of $(x_{t+1} - x_t)$ become meaningful as it indicates (x_{t+1}/x_t) approaches 1.

As it turns out in fact, the slow-rising initial section of the sigmoid curve is a mere consequence of the mathematical attribute of the population as a non-negative entity: the t-axis (at which $x_t = 0$ for all t) is the lowest physical bound, meaning that the lower section of the curve is compressed against the t-axis. In other words, the visual impression of an apparently slow start of the sigmoid growth curve turns out to be of no ecological significance. This is why the per-unit time rate of increase apparently contradicts the per-capita rate that exhibits the highest value at the start. [Note: As a matter of fact, the difference $(x_{t+1} - x_t)$ does carry the information if we know their exact values by calculation: it is just that we would not see it clearly on the graph.]

How about the ecological significance of the peak value of the difference $(x_{t+1} - x_t)$? Is there any? As we have seen in the foregoing table, the difference peaks at $t = 4$. This corresponds to the segment (x_4, x_5) of the curve in Figure 4.2c that exhibits the steepest ascent. This means that, at the segment, the sigmoid curve changes its rate of increase from high (accelerating) to low (decelerating). This segment (a 'point' if the curve is smooth as in a continuous-time process) is usually referred to as the 'inflection point' of the sigmoid curve. But, obviously, the inflection point is a physically inevitable consequence of the sigmoid curve whose starting section is physically compressed against the t-axis. It follows that neither does the inflection point carry an ecological significance.

After all, the iconic sigmoid curve of the classical logistic model hides the per-capita rate of change (the fundamental mechanism of population growth) at the start of population growth: it reveals the ecological mechanism only near the equilibrium level. Is there an alternative way to get around the problem? Yes, a logarithmic transformation of the population size x as well as the reproduction curve, as I now show.

4.6 Logarithmic Transformation of a Reproduction Curve

Here I use model (4.1), repeated below, as a simple and convenient example:

$$r_t = r_m \exp(-cx_t). \qquad (4.1 \text{ rpt})$$

After the (natural) logarithmic transformation (henceforth ln-transformation), we have:

$$\ln(r_t) = \ln(r_m) - cx_t. \qquad (4.4a)$$

[Notes: In the above transformation, I used the rules: $\ln(uv) \equiv \ln(u) + \ln(v)$ and $\ln[\exp(u)] \equiv u$. Incidentally, I would use the natural logarithm

4.6 Logarithmic Transformation of a Curve

'ln' in analysing a mathematical model as it is convenient in algebraic manipulations. In the following, I use the space-saving notations $\ln(x_t) \equiv X_t$ and $\ln(r_t) \equiv R_t$. However, when working on a graphical representation of an observed series, the common '$\log_{10}$' would be much more natural to use. Whenever I see no need of differentiating between the ln- and $\log_{10}$-transformations, I may use the expression 'log-transformation'.] Thus, equation (4.4a) can be written as:

$$R_t = R_m - c \exp(X_t). \qquad (4.4b)$$

[Note: $x \equiv \exp(X)$ is the inverse (i.e. anti-ln transformation) of $\ln(x) \equiv X$. It might look silly to ln-transform it first and, immediately after that, to anti-ln it, but it is necessary to maintain the consistency in using the notation X instead of the problematic expression $x \geq 0$.]

Now, let us draw a reproduction curve generated by the model (4.4b), i.e. the R_t-vs.-X_t version of the original (x_{t+1}-vs.-x_t) reproduction curve (or the Moran plot) in Figure 4.2a,b.

4.6.1 Drawing the Reproduction Curve after ln-Transformation

Let us do it first by a numerical calculation with the (arbitrarily chosen) constants $(r_m, c) = (2.5, 1.0)$ as in Figure 4.2, such that $R_m = \ln(r_m = 2.5) \approx 0.9163$. Set a series of numbers for X_t (in one column of a spreadsheet), ranging from -3 to 0.5 with 0.1 intervals, and let the computer calculate R_t for each value of X_t by formula (4.4b). Now, create a new graph (Figure 4.3a), and plot the computed R_t on the vertical axis (with the tick marks ranging from -1.0 to $+1.0$) against X_t on the horizontal axis.

In the graph, place the R_t-axis at the left end of the X_t-axis so that it would not visually interfere with the curve. Make sure the physical length of a unit interval is identical between the two axes as in Figure 4.2b: we will use this feature shortly. The placement of the X_t-axis with respect to a given tick mark of the R_t-axis is arbitrary, but there should be a horizontal axis at $R_t = 0$: call it the ($R_t = 0$)-axis, which corresponds to the 45° line in Figure 4.2a and b. Then, the intersection between the (log-transformed) reproduction curve and the ($R_t = 0$)-axis gives the equilibrium point X^*, i.e. the ln-transformed x^* of Figure 4.2a and b.

Now that R_m is the maximum possible value of R_t in model (4.4b), the R_t-X_t reproduction curve stays below the dashed horizontal line at $R_m \approx 0.9163$. Also, noting that $R_t = 0$ at the equilibrium level, substituting $R_t = 0$ in (4.4b), and solving for $X_t = X^*$, we find:

70 · Reproduction Curves and Their Utilities

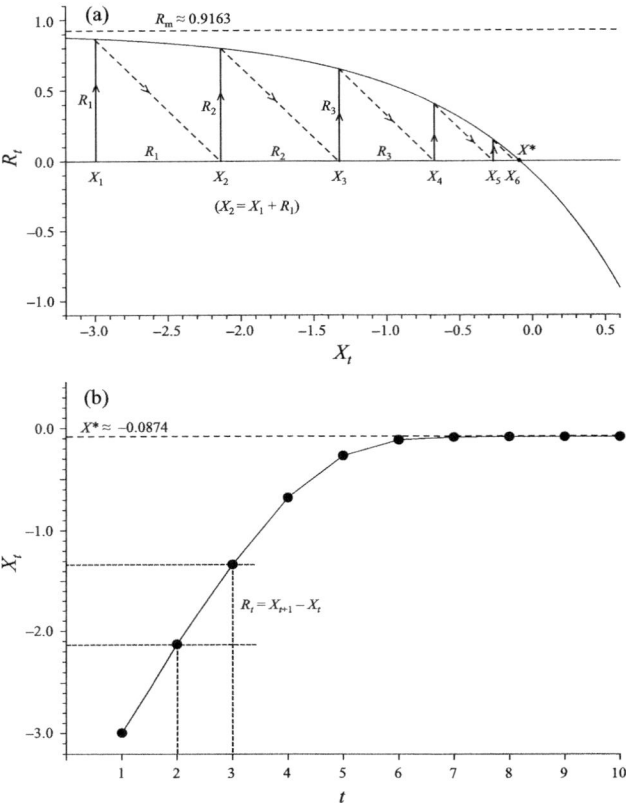

Figure 4.3 (a) A log-transformed reproduction curve generated by model (4.4b), and the graphical method of generating a population series. (b) The series $\{X_t = \ln(x_t)\}$ generated thus.

$$X^* = \ln(R_m/c). \quad (4.5)$$

For $R_m = \ln(2.5) \approx 0.9163$ and $c = 1$, we find $X^* \approx \ln[\ln(2.5)] \approx -0.0874$. We are now ready to graphically generate the population series $\{X_t\}$.

Let us start with an arbitrary initial point $X_1 = -3$, say, on the ($R_t = 0$)-axis. From this point, draw a vertical line (with an arrow) upwards until it hits the curve. The height of the line gives R_1. Now, from where the line hits the curve, draw a (dashed) line to the right and downwards at an angle of 45° until it intersects the ($R_t = 0$)-axis. The distance between X_1 and the intersection is equal to R_1 so that the intersection reads X_2: this is because the definition $r_t = x_{t+1}/x_t$ has been transformed to

$R_t = X_{t+1} - X_t$, and hence $X_1 + R_1 = X_2$, remembering that the length of the unit interval has been equalized between the R_t- and X_t-axes.

Repeating the above process, you can generate the series $\{X_t\}$ for $t = 1, 2, 3, \ldots$ to find it gradually and monotonically converges to X^*. We see that, as compared with its pre-transformed counterpart in Figure 4.2b, the R–X reproduction curve here is much simpler to use for generating $\{X_t\}$ graphically. But, most importantly, it clearly shows that the (per-capita) reproductive rate R_t is highest at $t = 1$, i.e. R_1 is the highest at the point of start, i.e. X_1, and decreases steadily and consistently as X_t increases.

4.6.2 Time Plot of the ln-Transformed Population Series

We now plot the (log-transformed) series $\{X_t\}$ against t in Figure 4.3b and compare it with the non-negative series $\{x_t\}$ in Figure 4.2c. We see that the series $\{X_t\}$ is no longer sigmoidal, even though it is mathematically equivalent to the sigmoidal series $\{x_t\}$. More importantly, the log-transformed series $\{X_t\}$ readily depicts the per-capita rate of population increase: unlike the difference $(x_{t+1} - x_t)$ in Figure 4.2c, the distance between two adjacent points, i.e. $(X_{t+1} - X_t)$, of the series $\{X_t\}$ in Figure 4.3b is equal to $R_t = \ln(x_{t+1}/x_t)$, i.e. ln(per-capita rate of change). Indeed, the difference $(X_{t+1} - X_t) = R_t$ in Figure 4.3b is highest at the start and steadily decreases as X_t approaches X^*. So, we have got rid of the nuisance physical constraint: the population size (x) is non-negative.

Altogether, the logarithmic transformation is like getting two ducks in one shot. Not only is the R–X reproduction curve in Figure 4.3a is much easier to use in generating the series $\{X_t\}$, but also the time plot of $\{X_t\}$ in Figure 4.3b readily depicts the per-capita rate of change R_t. Thus, for the analysis of the time process of populations, the log-transformation is the way to go. Do not get me wrong, though: I do not mean that a non-negative measure of population size is no good. Far from it! We would definitely need it in theoretical investigations, e.g. when building mathematical models as in Chapter 3. So, you should selectively use whichever format is appropriate in your own work.

4.7 An Application to Actual Data

Using the foregoing R–X representation of the reproduction curve, let us take a fresh look at the US census data (from 1790 to 1910 at 10-year intervals) to which Raymond Pearl and Lowell Reed (1920) fitted their 'logistic' model, their equation (ix), which is equivalent to the common-version model (3.3c) in Chapter 3. [Note: Their equation (ix) is characterized by three parameters (a, b, c) but is of the same form as (3.3c) in Chapter 3 with the notational equivalence: $(\rho, K) \equiv (a, b/c)$.]

Pearl and Reed presented the numerical data of the census results up to 1910 in table 1 of their 1920 article, and fitted their equation (ix) to the data in their figure 3. I reproduce their curve here in Figure 4.4a (solid curve), using their equation (xviii). Furthermore, I extended the census data up to 2000, using the information provided in 'Chart of US Populations, 1790–2000' online from census-charts.com.

In Figure 4.4a, the series of solid circles is the actual census data up to 2000. The solid smooth curve is calculated by Pearl and Reed up to 1910 with my extension to 2000, using their predictive equation (xviii). I also added the dotted curve of the Malthusian exponential progression for comparison. [Note: I calculated the Malthusian curve by the formula $x_t = x_1 r_m^{t-1}$ in which t (= 1, 2, 3, ...) represents the t-th decade (beginning with 1790 as the 1st) and $r_m = x_2/x_1 = 5{,}308{,}483/3{,}929{,}214 \approx 1.3510$.] We see that the three curves are visually indistinguishable until about 1850. From then on the Malthusian took off, while the census curve was indistinguishable from the solid smooth curve (fitted by Pearl and Reed with my extension) for another century until 1950, but the actual population kept increasing thereafter.

In Figure 4.4b, I re-plotted the above three curves in the common log-scale. We see that the fitted (solid) curve exhibits no sigmoidality, but only increases with a steadily decreasing rate over the period up to 1950, much the same manner as Figure 4.3b exhibits. [Note: The log-transformed Malthusian is a straight (dotted) line because its rate of increase is constant at $R_m = \log(r_m = x_2/x_1) = \log_{10}(1.351) \approx 0.131$, as noted in the preceding paragraph.]

In Figure 4.4c, I transformed the time-plot representation of Figure 4.4b into the R–X reproduction curve, comparable to Figure 4.3a. [Note: Unlike in Figure 4.3a, the scale of the vertical axis in Figure 4.4c is not equalized to that of the horizontal axis but expanded so as to make the trend clearly visible. This modification is allowed here because the curve is not intended to be used for generating the population series.] The resultant reproduction curve brings out details of what went on in US population changes that are not clearly visible in the time-plot representation (either in graph a or in graph b).

There are two issues that Figure 4.4c reveals. First, the (per-capita) rates of change in the US population over these decades were the highest

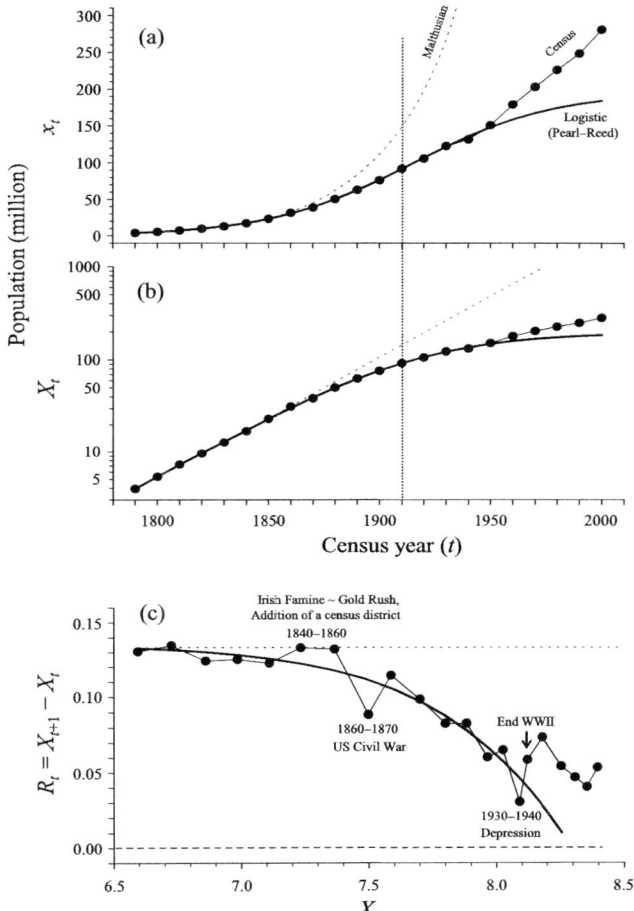

Figure 4.4 Graphical representation of the US census data (solid circles), fitted with the Pearl–Reed equation of the 'logistic' curve up to 1910 with my extension to 2000 (solid curve). (a) In non-negative linear scale on the vertical axis. (b) The common log-transformation of (a). (c) Transformation of (b) into a reproduction curve in the R–X format.

at the beginning of the census period and steadily decreased, following the Pearl–Reed 'logistic' prediction quite well up to the end of the World War II. Second, the observed $\log_{10}$-reproductive rate R_t (connected solid circles) exhibits considerable jitters (about the 'logistic' prediction curve), as compared with the much smoother time-plot representations in Figure 4.4a,b. The most noticeable feature in

74 · Reproduction Curves and Their Utilities

Figure 4.4c is a conspicuous dip in the observed rate of change (relative to the prediction curve) over the decade from 1860 to 1870. This dip coincided with the civil war of the country during the early half of the decade. Also noticeable is another dip in the 1940 census which coincided with the period of the Great Depression during the early 1930s. Although barely noticeable in the solid-circle series in Figure 4.4a, the census-charts.com people did notice these dips in their time-plot chart of the census data (they must have had a very good eye). They rightly attributed the dips to substantial decreases in immigration from other countries during those decades.

However, my graph in Figure 4.4c shows noticeably high rates of change during the two decades from 1840 to 1860: more precisely, the first high is the rate of change from the 1840 census to the 1850 one, followed by the rate between the 1850 and 1860 censuses. Even the keen-eyed census-charts.com people missed these highs as they are not noticeable in the time-plot of populations in Figure 4.4a: evidently the linearly scaled curve at a low level hid them. Neither are these high rates noticeable even after the log-transformation in Figure 4.4b because the resolution in the time-plotted curve is not high enough visually. However, these are clearly brought out in the R–X representation of the reproduction curve in graph c. I am almost certain that these highs in Figure 4.4c are attributable to the Great Famine in Ireland that occurred between 1845 and 1849 (the effect of which is manifest in the 1850 census), immediately followed by the Gold Rush in California from 1848 to 1855 (manifest in 1850 and 1860 censuses). No doubt, these two independent but consecutive events brought in a substantial number of immigrants to America. Besides, and very importantly, no census had been made in the region 'West' (designated by the US Census Bureau) prior to the 1850 census which included the region for the first time. This must have also contributed to a substantial rise in the 'total population' at the 1850 census year. After all, the raw census data (that are used in these graphs) are not in terms of population density but the total population of the country which involves the spatial expansion of the population, the aspect not taken into account in the Pearl–Reed ('logistic') model. And then, of course, the most noticeable deviations of the census data from the Pearl–Reed prediction occurred in 1950 and onwards: the deviations are, needless to say, attributable to a substantial rise in immigration soon after the end of World War II in 1945.

Altogether, the effect of human migrations offsets the Pearl–Reed prediction of the US population growth. This is inevitable because their model (as well as the Verhulst logistic model) was not built to accommodate such external factors (or exogenous processes). Was the logistic model no good? Do not make a hasty judgement: the value of a good model is to provide a criterion for detecting what causes the observation deviating from the model. This recognition leads me to a generalization of model (4.4b) so as to accommodate the effect of such exogenous processes as immigrations. I will come back to this issue later in the present chapter. For now, I suggest that we look into the varieties of dynamical (time-dependent) pattern that model (4.4b) in its present form (not taking account of exogenous influences) can generate.

4.8 Variation in Dynamical Pattern of the Model Process (4.4b)

Recall that, when drawing the R–X reproduction curve in Figure 4.3, I used a particular set of parameter values: $(R_m, c) = (0.9163, 1.0)$. Let's see what happens if we change R_m to 1.9, say, while keeping $c = 1.0$ as before. Figure 4.5a shows the reproduction curve computed with this new set of parameter values. The curve intersects the $(R_t = 0)$-axis at $X^* = \ln(1.9/1.0) \approx 0.64$ by formula (4.5). Using this reproduction curve, let us graphically generate the population series $\{X_t\}$, following the procedure in Figure 4.3a.

Let us start in Figure 4.5a with an arbitrary initial population size $X_1 = -1.5$ below (left of) X^*. We find that X_2 is still below X^* but X_3 overshoots X^* where the corresponding R_3 becomes negative, such that X_4 goes back to below X^*. And the process in Figure 4.5a goes in cycles about X^*, although gradually converging to it. The time plot of the series $\{X_t\}$ in Figure 4.5b jumps up from X_1 to X_2; over shoots the equilibrium (X^*) level at X_3; thereafter oscillates (in a saw-toothed manner) about the X^* level; but eventually converges to the equilibrium level at $R_t = 0$. However, this particular feature would change again if we increase the R_m value even further. But, before we get to that issue, some more aspects of the mathematical characteristics of model process (4.4b) should be understood: in particular, the role played by parameter R_m in determining the dynamical pattern of the series $\{X_t\}$.

4.8.1 Dependence of the Dynamical Pattern on the Parameter R_m

By now, you might have noticed that the pattern of the series $\{X_t\}$ is dictated by the slope (tangent) of the reproduction curve at X^*. You can

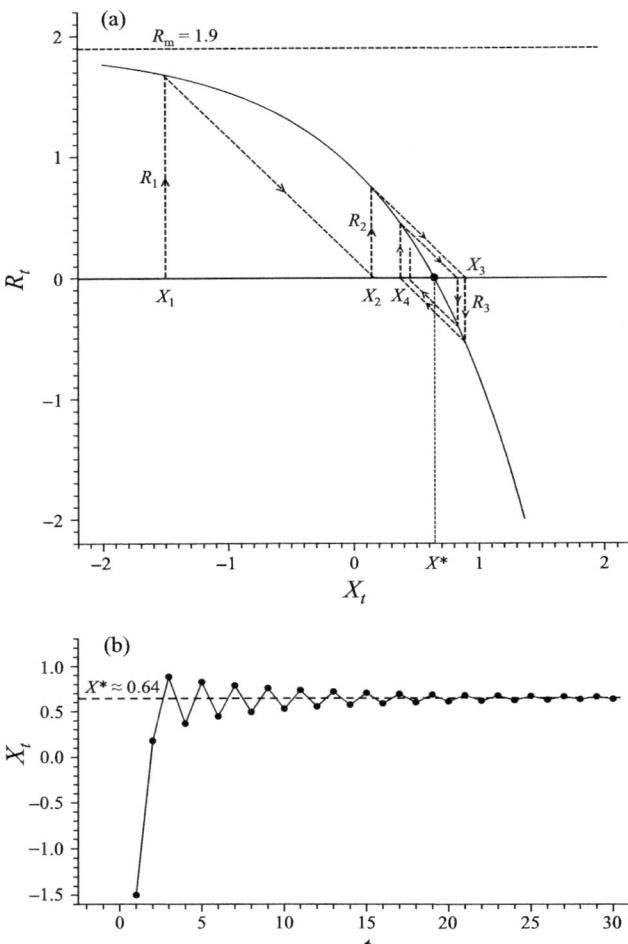

Figure 4.5 (a) Reproduction curve similar to the one in Figure 4.3a, but with the R_m value raised to 1.9. (b) The resultant population series exhibiting saw-tooth oscillations.

guess from the foregoing examples the following tendencies. First, the series is smoothly and asymptotically convergent to X^* if the slope is no steeper than $-45°$ as in Figure 4.3a, remembering that a negative slope indicates a downward slope. Second, if the slope is steeper than $-45°$ as in Figure 4.5a, the series becomes cyclic (oscillatory) about X^*. But, what determines the slope at X^*? In model (4.4b), it is the parameter R_m exclusively for the following reason.

As has been explained in detail in Appendix 4A, the slope (tan θ) of a curve, e.g. a generic function $f(u)$ plotted against u, is shown to be measured by the derivative $df(u)/du \equiv f'(u)$. In the present model (4.4b), $u \equiv X_t$ and $f(u) \equiv R_t$. Thus, $\tan\theta = dR_t/dX_t$. This derivative can be readily found using the list of the derivatives of certain standard functions given in Appendix 4B. In particular, by the case (i) in the list, the derivative of the first term (i.e. R_m) on the right-hand side of model (4.4b) is 0 because R_m is constant. Also, by the special case of (iv), the derivative of the second term, i.e. $d[-c\exp(X_t)]/dX_t$, is equal to $[-c\exp(X_t)]$: that is, the derivative of the second term is equal to itself. Thus, we find:

$$dR_t/dX_t = -c\exp(X_t). \tag{4.6}$$

In the meantime, we know $X^* = \ln(R_m/c)$ by (4.5). So, the slope of the curve at X^* is given by substituting $X^* = \ln(R_m/c)$ for X_t on the right-hand side of (4.6). That is, noting that $\exp[\ln(u)] = u$, we find the derivative dR_t/dX_t at $X_t = X^*$ (appended by the notation '$|_{X^*}$') to be:

$$dR_t/dX_t|_{X^*} = -c\exp[\ln(R_m/c)] = -c(R_m/c) = -R_m. \tag{4.7}$$

We see that the slope of the reproduction curve of model (4.4b) at X^* depends solely on R_m.

This attribute is in fact peculiar to the present model (4.4b); I will talk about its generalization later in Chapters 5 and 6. Nonetheless, it is informative if we know why this happens to model (4.4b). The following geometry of reproduction curves generated by (4.4b) would let you see 'why' in a straightforward manner.

4.8.2 Geometry of Reproduction Curves Generated by Model (4.4b)

First of all, as it so happens, the model generates one and only one basic shape of reproduction curve (the prototype, as it were) that does not depend on the parameters R_m and c. That is, all curves that model (4.4b) generates in the (R_t, X_t) coordinate system are congruent, regardless of the value of R_m or c. Only the position of a curve in the coordinate system (with the horizontal axis fixed at $R_t = 0$) is determined by (dependent on) the parameters involved. In particular, the vertical position is determined solely by R_m, whereas the lateral position, solely by c.

To confirm these attributes, compare the curve in Figure 4.3a and the other in Figure 4.5a. The only difference you would see between the two is that the curve in Figure 4.5a is placed higher than the one in Figure 4.3a. But $R_m = 1.9$ in Figure 4.5a and $R_m \approx 0.9163$ in

Figure 4.3a. So, if you bring down the Figure 4.5a curve vertically as much as $1.9 - 0.9163 \approx 0.9837$, and after a little jiggling to adjust for the approximation, it should precisely match the Figure 4.3a curve.

It is important to notice that as a given curve is shifted up or down vertically in relation to the $R_t = 0$ axis, the slope of the curve at X^* changes accordingly: an upwards shift makes the slope steeper and the other way in a downwards shift. Also evident is that an upwards shift moves the position of X^* to the right, whereas a downwards shift moves it to the left.

What if we change the value of parameter c? To find an answer, let us write $c \equiv \exp(C)$. Then, the expression '$c \exp(X_t)$' in (4.4b) can be written as $\exp(C) \times \exp(X_t) \equiv \exp(C + X_t)$. Thus, (4.4b) is written as:

$$R_t = R_m - \exp(C + X_t). \tag{4.8}$$

Now, while keeping the R_m-value fixed, we reduce the c-value from 1 (or $C = 0$) in Figure 4.5 to about 0.607 such that $C = -0.5$. This time you see the curve shifted laterally to the right by as much as 0.5. That is, the equilibrium point X^* moves with the curve to the right as much as 0.5 units: conversely, you can readily verify that an increase in C shifts the curve (and hence X^*) laterally to the left. It is important to notice here that a lateral shift of the curve does not change its slope at X^*, such that the resultant time-plot of $\{X_t\}$ remains unchanged in pattern about the (laterally shifted) X^*. Appendix 4D explains the mathematics of the above attributes exhibited by model (4.4b).

To summarize, in (4.4b), parameter R_m changes the vertical position of the reproduction curve and c, the lateral position. However, no matter how these values change, the basic shape of the curve is invariant. Thus, as R_m becomes high (low), the curve is lifted up (brought down); its slope at X^* becomes steeper (or less steep); and the pattern of the series $\{X_t\}$ about X^* changes accordingly. I now show how it changes systematically by examples.

4.9 Examples of Variations in Dynamical Pattern

In Figure 4.6, I generated typical examples from four major classes (according to the variation in R_m and, hence, the slope of the reproduction curve at X^*) of dynamical pattern. To make them directly comparable with each other, the X^* level is arbitrarily fixed and standardized at 0 by adjusting the C-value in (4.8). [Note: I chose $X^* = 0$ conveniently because, after all, the dynamical pattern of the series $\{X_t\}$ depends only on the slope of the reproduction curve at X^* and, as already shown in

4.9 Variations in Dynamical Pattern

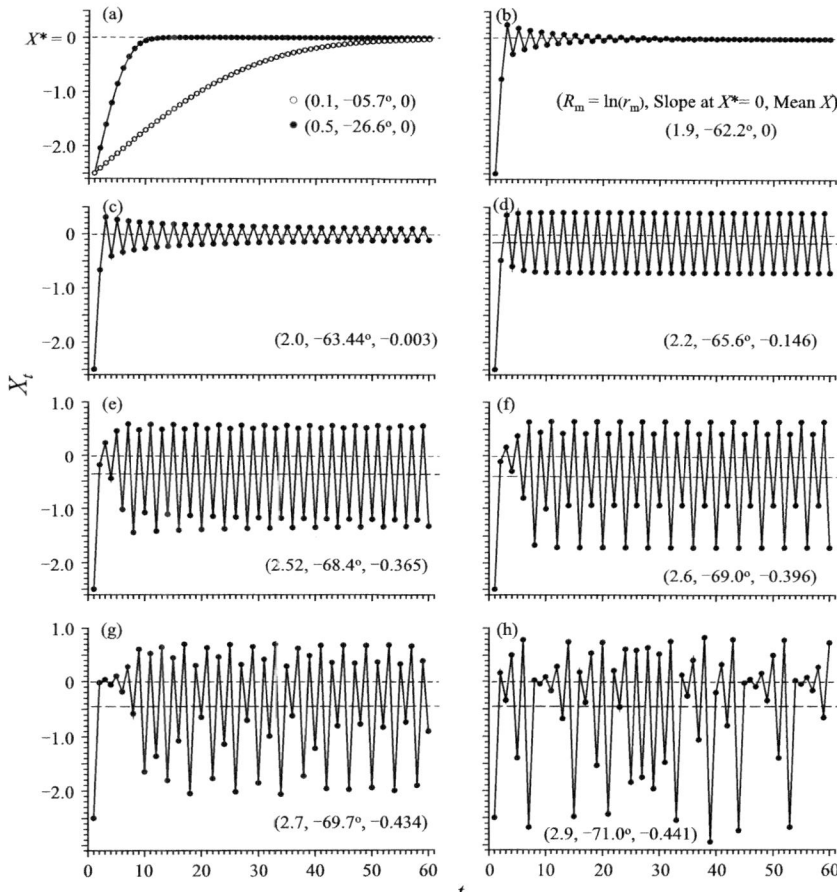

Figure 4.6 Examples of dynamical pattern in population that model (4.4b) generates with different R_m values.

(4.7), the slope is equal to $(-R_m)$ no matter what the actual value of X^* might be.] Also shown in each graph is the mean level of the series $\{X_t\}$, using the section of the series well after it has reached the equilibrium level at $X^* = 0$. [I calculated the mean of $\{X_t\}$ after generating a series of 550 points and knocking off the first 50 points to get rid of the effect of an initial non-equilibrium section.] As will be explained shortly, the mean level of the series $\{X_t\}$ generally differs from the equilibrium level X^*. To make the difference explicit, the X^* level is marked with a short-dashed line, compared with the mean level, marked with a longer-dashed line.

The graphs are arranged in the order of increasing R_m. In each graph, the value of R_m, the corresponding slope in degrees, i.e. $\tan^{-1}(-R_m)$, of the reproduction curve at X^*, as well as the mean of $\{X_t\}$ are shown in parentheses. The designation of each value is listed in graph b. Now the details.

Class of monotonic convergence to X^*. In this class, the slope of the reproduction curve at X^*, i.e. $\tan^{-1}(-R_m)$, is no steeper than $-45°$, as exemplified by the two examples in Figure 4.6a: one for $R_m = 0.1$ and the other for 0.5, or in general $0 < R_m \leq 1.0$. Both series monotonically and smoothly converge to X^* as t increases. However, the series for $R_m = 0.5$ reaches the X^* level at a much faster rate, as compared with the one for $R_m = 0.1$. Because both curves are monotonically convergent to X^*, the mean of each series converges in the long run (or asymptotically) to $X^* = 0$.

Class of convergent oscillations. The slope in this class is in the range $45° < \tan^{-1}(-R_m) < 63.4°$ in absolute value, corresponding to $1 < R_m < 2$. The dynamical pattern in this class is exemplified by Figure 4.6b, in which the series $\{X_t\}$ exhibits saw-toothed oscillations, yet it converges to X^*. Because the series is convergent to X^*, its mean converges to X^* in the long run.

In Figure 4.6c, R_m is increased to 2.0, and the slope $\tan^{-1}(-R_m)$ becomes a little steeper than $-63.43°$. The series $\{X_t\}$ maintains the saw-toothed oscillations of graph (b), but no longer converges to X^* exactly within the finite number of time steps: it converges only in the limit ($t \to \infty$). We see that, in the example, even though oscillating about the equilibrium level for quite a long time, its realized mean is -0.003, not quite convergent to 0, albeit 0 practically. In fact, this is a transition from the class of convergent oscillations to the class of limit cycles, as shown below.

Class of limit cycles. In Figure 4.6d with R_m increased to 2.2, i.e. $\tan^{-1}(-R_m) \approx -65.6°$, the series $\{X_t\}$ exhibits a perpetually fixed oscillatory pattern, called a limit cycle: so called because, no matter where it is started, unless exactly at X^*, the series $\{X_t\}$ would converge (in the limit $t \to \infty$) to a fixed cycle. In this example, starting at $X_1 = -2.5$, it has practically reached a limit cycle by $t = 10$.

Another attribute of the series in Figure 4.6d that is important in the present context is the following. The mean of the series would noticeably deviate downwards from the equilibrium level at X^* by as much as -0.146. This is because the reproduction curve is asymmetric (horizontally) about the X^*, and so the series $\{X_t\}$ oscillates asymmetrically about the X^* level: a downwards deviation is greater than an upwards deviation in absolute value.

A further increase in R_m to 2.6, i.e. $\tan^{-1}(-R_m) \approx -69°$, in Figure 4.6f, the $\{X_t\}$ exhibits a complex pattern, although still within the class of limit cycles. Nonetheless, the maximum amplitude of the cycle is much greater than that of the series in Figure 4.6d. Accordingly, the deviation of the mean from X^* in Figure 4.6f is now quite substantial. Figure 4.6e exhibits a transitional pattern from Figure 4.6d to Figure 4.6f.

Class of deterministic chaos. Finally, somewhere around $R_m = 2.7$ or $\tan^{-1}(-R_m) \approx -69.7$ as in Figure 4.6g, the cycle begins to exhibit a sign of irregularity. With $R_m = 2.9$, or $\tan^{-1}(-R_m) \approx -71°$ in Figure 4.6h, the series $\{X_t\}$ no longer maintains a fixed pattern but exhibits an unpredictably complex pattern, the so-called 'deterministic chaos' in applied mathematics. But it is important to notice that, as compared with the previous class of limit cycles, the amplitude of the chaotic oscillations is even greater, resulting in the observed mean of the series deviating even more from the equilibrium level at X^*. [Note: Albeit irregular, the chaotic process is not a random process, in that it is totally determined, given the parameter values and the initial state X_1, hence the name 'deterministic chaos'.]

In all graphs of Figure 4.6, the population series $\{X_t\}$ started below the equilibrium level X^*. What if started from above it?

The series $\{X_t\}$ starting above X^*. There are two situations to consider according to the slope of the reproduction curve at X^*, say θ^*: (i) $0 > \theta^* > -45°$ as in Figure 4.6a; (ii) $\theta^* < -45°$ as in all other graphs in Figure 4.6.

Figure 4.7 illustrates situation (i) in which the $-45°$ (dashed) reference line intersects the curve below the horizontal $(R_t = 0)$-axis. The X-coordinate of the intersection is marked as X_s on the $(R_t = 0)$-axis. Now, the graph is divided into two sub-situations according to the position of X_1 relative to X_s: sub-situation (ia) X_1 is placed between X^* and X_s, and (ib) above (right to) X_s. We see that in (ia) the subsequent series $\{X_t; t = 2, 3, 4, \ldots\}$ converges directly (asymptotically) to X^* from above, whereas in (ib) X_2 plunges (to the left) below X^* but the subsequent series $\{X_t; t = 3, 4, \ldots\}$ converges asymptotically to X^* as in Figure 4.6a.

In situation (ii), although not illustrated, we can readily see that the $-45°$ reference line never intersects the reproduction curve below the $(R_t = 0)$ level. Consequently, X_2 (following X_1 placed above X^*) plunges to below (the left of) X^*, in much the same way as in the (ib) situation in

82 · Reproduction Curves and Their Utilities

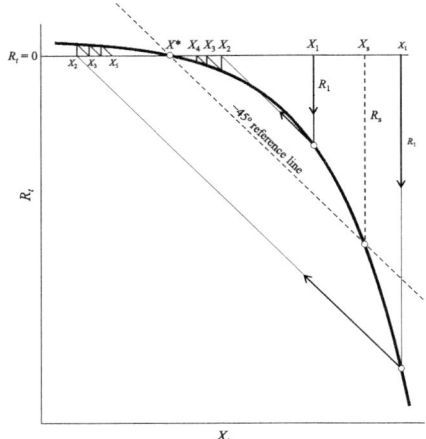

Figure 4.7 Graphical consideration of the situation in which the population starts above the equilibrium X^*.

Figure 4.7. From then on, the series $\{X_t;\ t = 3, 4, \ldots\}$ repeats the pattern as in any one of the graphs b to f in Figure 4.6 according to the value of R_m. You can readily confirm these patterns by simulations.

Summary of section 4.9. The series $\{X_t\}$ that model (4.4b) generates exhibit their dynamical patterns (about X^*) in four major classes. In the order of increasing R_m, these are: monotonic convergence, oscillatory convergence, limit cycles, and finally chaos (or 'chaotic cycles'), with transitional patterns in between. The pattern of a given series at the starting section depends on the initial position of X_1 relative to X^*.

Now, recall that model (4.4b) is the ln-transformed version of (3.10), which is a theoretical version of the classical logistic models (3.1a, b, or c) in Chapter 3. But, the following difference is noticeable: while (4.4b) generates a variety of dynamical patterns, the classical model generates only one pattern, i.e. asymptotic convergence to its equilibrium level, e.g. K in the common version (3.1c) for $\rho > 0$. Why the difference? Basically, it is due to the fact that the classical model is a continuous-time process, whereas model (4.4b) is discrete in time. The detail is as follows.

4.10 Difference between Discrete-Time and Continuous-Time Processes

The following analogy would hopefully let you see the difference in an intuitive manner. An equilibrium level, be it K in the (common-version) classical model (3.1c) or X^* in the theoretical model (4.4b), acts as a border wall separating the coordinate space into two sections: below and above it. In process (4.4b), its path between the two consecutive generations is discrete: x_t does not change continually to x_{t+1} but jumps to it. Therefore, as it approaches the wall (of equilibrium level) from one side, it can jump over to the other side like a hare; it may decide not to jump, though. This hare-like athleticism of the discrete-time process generates various patterns as we see in Figure 4.6. On the other hand, the classical model as a continuous-time process is like the tortoise: it cannot jump over the wall (of equilibrium) because its pace of walk is as small as the differential dx (i.e. infinitesimally small). It only can walk up slowly to the wall from one side. However, once it has reached the wall, it would only be able to walk along the wall thereafter, unless you picked it up and put it over to the other side: then, it will approach the wall again slowly and asymptotically from that side. Altogether, the discrete-time model is much more flexible than the continuous-time model in generating the dynamical patterns as we often observe in nature.

Now, although the discrete-time model (4.4b) can generate various dynamical patterns, not all of the varieties are ecologically feasible. In particular, the feasibility tends to diminish as the (ln) potential reproductive rate R_m increases, as I discuss below.

4.11 Ecological Feasibility of Variations in Discrete-Time Processes

Consider those processes for high R_m values as exemplified in Figure 4.6e–h, and visualize the reproduction curve that generated one of them: that is, visualize one like the curve in Figure 4.5 but with a much higher level of R_m and with an extension towards lower values of X_t, while X^* is still fixed at 0. What we would see is the following.

If placed initially (as X_1) much below X^*, the population would approach X^* consistently at the rate R_t only slightly less than R_m. But, when X_t comes close to X^*, the rate R_t suddenly becomes very sensitive to X_t, indicated by a very steep (downwards) slope of the curve about X^*. As a result, the production of offspring in the following generation (X_{t+1}) may plunge to an extreme low level which, in turn, lowers the mean level of the population over a number of generations. In Figure 4.6e–h, the mean levels are between −0.365 and −0.441, compared with the $X^* = 0$ level. Translating those back into the linear scale by anti-ln transformations, these mean levels are about 70%–65% of the $x^* = 1$ level, i.e. substantial reductions by 30%–35% from the equilibrium density. In contrast, for the lower values of R_m in Figure 4.6a–c, there is, practically, no reduction in the mean levels from X^*.

The foregoing attributes are at the population level. The crucial issue here is at the individual level. Consider those individuals with the genetic traits that: they are highly productive when just a few competing neighbours are around, but they become extremely (excessively) intolerant of the neighbours when the population approaches the X^* level. The average fitness of these individuals would likely be lower than the fitness of those that has a moderate potential reproductive rate and are more tolerant of the presence of their neighbours. It suggests that, in the long run, the individuals that are too competitive and aggressive would be selected out, and those that are moderate in temperament would prevail in the population. It follows that the likelihood of the occurrences of high-amplitude limit cycles and chaos would be rather low among natural populations. [Note: The above perception is logically valid within the framework of single-species processes. A predator–prey interaction process may more readily promote a limit cycle as I discuss in Chapter 8.]

The theory of deterministic chaos has stimulated applied mathematicians and ecologists alike since the early 1970s. However interesting it may be from the mathematics point of view, its likelihood of occurrence in an ecological process in its natural environment would, as I see it, be low or even doubtful. A caveat is this: any ecologically feasible process has to obey a mathematical (physical) possibility, whereas the converse is not true: a mathematical possibility does not necessarily imply ecological feasibility.

Now, what about a lower bound in R_m: would it exist? I think it would. Consider Figure 4.6a for instance. If R_m was very low, the population may reach the X^* only very slowly. Then, once it had become very low, the population may not recover quickly enough. That is, the risk of extinction would be high in natural conditions. At the individual level, the fitness of those with a low R_m would on average be low, and would be unable to compete with those individuals with a (moderately) higher R_m.

So, I suggest that the range of variations in the potential reproductive rate is subject to natural selection, which implies the existence of an optimal level, depending on the species life history. Leaving this issue open to discussion among the readers, I move on to another subject:

investigations into the effects of external factors on the dynamics of a population. To begin with, I introduce the notions of endogenous and exogenous processes.

4.12 Endogenous and Exogenous Processes

When constructing the theoretical model (3.10) in Chapter 3, I considered the process that was idealized to be a self-contained function of population density in a non-fluctuating environment. Thus, we may consider the basic model (4.1), and its ln-transformed variants (4.4b) or (4.8), to represent an endogenous process. Also, we may call r_m and c, or their variants $R_m \equiv \ln(r_m)$ and $C \equiv \ln(c)$, *endogenous* parameters in that they characterize the endogenous process. In contrast, we may call environmental influences 'exogenous processes', or 'exogenous influences'. But what effects do the exogenous influences exert on the endogenous process? They affect parameters R_m and C, as I now show with a particular reference to the US population process we examined earlier.

By 1790, the first census year, the Industrial Revolution was well under way. In particular, the advent of steam engines revolutionized transoceanic transportation, which promoted a steady increase in immigration into America. Furthermore, during the entire period from 1790 to the present, human populations have undergone enormous influences from technological development, especially in agricultural technology, which increased the level of resource supplies. In addition, better public hygiene and sophisticated medical technology, etc. have reduced mortality. These changes must have caused a continual increase in the potential reproductive rate, i.e. the parameter R_m.

In the meantime, technological advancement also resulted in substantial changes in many aspects of daily life, which must have influenced the parameter C, originally defined as a measure of the intensity of within-population competition: in particular, a competitive action increases the value of C, whereas a cooperative (social) activity reduces it. Thus, it is most likely that, on the whole, an improvement in the quality of life has reduced it.

Now, these exogenous influences can be incorporated into the endogenous model (4.8) in the following way.

4.12.1 Reproduction Curve Under Exogenous Influences

Let ε_t (the lower-case Greek letter 'epsilon') be the effect of exogenous influences on R_m during generation t, such that the sum $(R_m + \varepsilon_t)$

replaces the plain (endogenous) R_m on the right-hand side of (4.8). Likewise, parameter C can be influenced by ζ (the lower-case Greek 'zeta') such that C can be replaced by $(C + \zeta_t)$. Thus, (4.8) is generalized to an endogenous–exogenous process:

$$R_t = (R_m + \varepsilon_t) - \exp[(C + \zeta_t) + X_t]. \qquad (4.9)$$

But what would the resultant reproduction curve look like? To find it in a simplified manner, let us consider the effect ε_t only, setting aside the effect ζ_t for the moment. The simulation in Figure 4.8a shows the series $\{\varepsilon_t\}$ superimposed upon the endogenous curve taken from Figure 4.3. We see the data point at generation $t = i$ deviates vertically from the endogenous curve as much as ε_i: for instance, $\varepsilon_3 = -0.12$, $\varepsilon_4 = 0.08$, ..., and so on. Now, follow the same procedure as in Figure 4.3, but instead of drawing a vertical line between X_i on the $(R_t = 0)$-axis and the endogenous curve, we draw a line between the axis and the data point (dot) that corresponds to X_i as illustrated. So, starting with (an arbitrarily chosen) X_1, we can generate the subsequent X_2, X_3, ..., and so on. We see that, in this particular example, the dots eventually congregate (albeit moving around continually) about the equilibrium point X^*. The resultant series $\{X_t\}$ is shown in Figure 4.8b.

Now, assuming that the $\{X_t\}$ in Figure 4.8b is an actually observed series, let us consider if we can find the endogenous form of reproduction curve hidden under exogenous influences.

4.12.2 Finding Endogenous Reproduction Curve Under Exogenous Influences

The above graphical exercise implies the following. Given an observed series $\{X_t\}$, we can, in theory, estimate the unknown (hidden) endogenous curve by plotting the difference $X_{i+1} - X_i$ (as the realization of R_i, $i = 1, 2, 3, \ldots, t$) against X_i, as illustrated in Figure 4.8a. Nonetheless, there are limitations in the above suggestion. First, the method works *provided* that the mean and variance of the series $\{\varepsilon_t\}$ remain unchanged during the observation period: evidently, the current climate changes would act against the proviso. Second, if the observed series was well regulated about the equilibrium point X^*, we would have just a blob of dots about X^* and the blob alone would provide little information about the curve: it might even give a distorted picture.

Then, would the above suggestion (of the method to find the hidden endogenous curve) be of no use? Probably 'yes' with many field

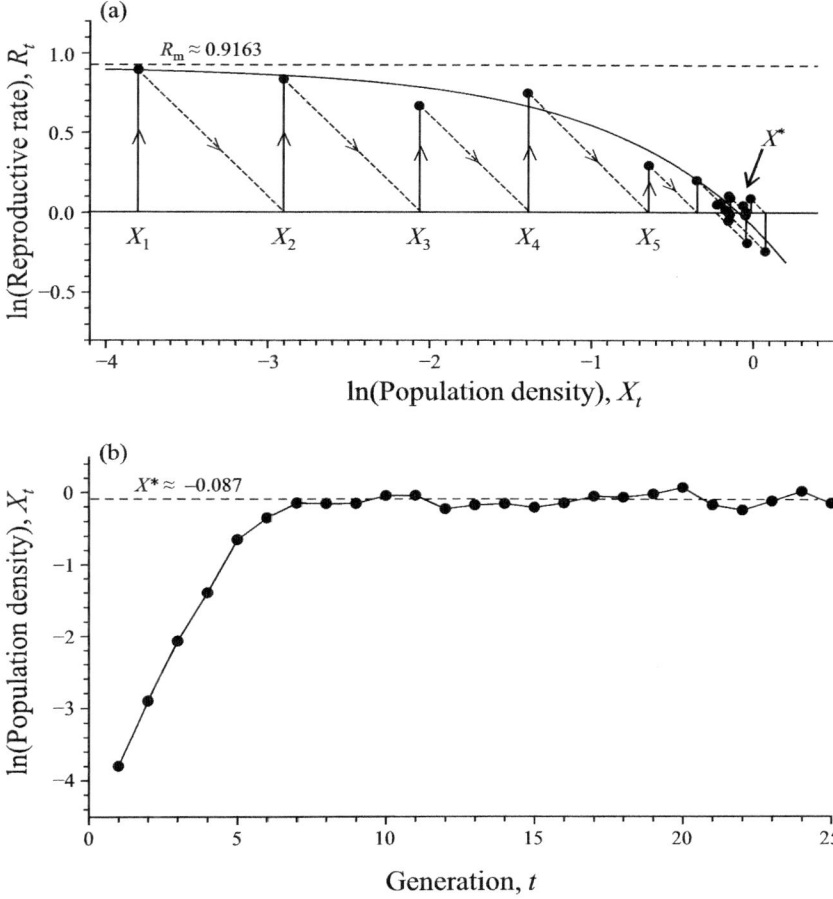

Figure 4.8 An idealized example of the effect of random exogenous influences incorporated into the endogenous curve in Figure 4.3.

observations conducted under uncontrolled (or more likely uncontrollable) natural environment. The real pragmatic value of the method is for designing and conducting experiments in a laboratory environment. For instance, by changing the initial density under controlled environmental conditions, it may be possible to expand the range of variation in the series such that we can readily see a major section of the endogenous curve.

Altogether, I strongly recommend that you do a lot of simulations with model (4.9) to get a feel for the probable effects of exogenous

88 · Reproduction Curves and Their Utilities

influences, including the random variate ζ_t acting on a lateral movement of the reproduction curve. You might even change the characteristics of the exogenous variate, e.g. give it a trend to imitate the effect of climate changes, etc. The knowledge you acquire in such exercises would be a good help for when you try to understand how and why an observed series behaves as it does.

4.12.3 Re-Assessment of the Pearl–Reed (Logistic) Model

At this moment, you might wonder why the classical Pearl–Reed model (commonly known as the 'logistic model' in the current ecology) as an endogenous model (not taking the effect of exogenous influences into account) did apparently fit the US population, at least until about 1950. Surely, as is evident in Figure 4.4c, there have been exogenous influences (the major one being immigration) from almost the beginning of the censuses. The most probable reason for the good fit is the following.

In the first place, the original Pearl–Reed model, i.e. equation (ix) in their original (1920) paper, has three constants, a, b, and c, to be estimated from data: cf. their differential equation (3.1b) in Chapter 3. So, they selected three points (1790, 1850, and 1910) in the actual census results, and the values of a, b, and c were so determined that the fitted curve would go through the results (data points) of these three census years. In other words, the Pearl–Reed estimates of a, b, and c were based on the 'average' of $(R_m + \varepsilon_t)$ and that of $(C + \zeta_t)$ over these decades. Therefore, it is not too surprising that the curve fitted well until 1910. It worked well perhaps because the effects of these exogenous influences on the US population were quite steady and consistent (that is, the mean and variance of the influences were more or less constant) over these decades or even into the subsequent decades up to 1950. However, the drastic changes in the rate of immigration that have occurred since the end of World War II (cf. Figure 4.4c) were such that parameters a, b, and c (estimated over the period prior to 1910) no longer fitted.

After all, it looks as though in the past two centuries, while the US population has been approaching its potential (endogenously induced) equilibrium level, the level has kept running away from it under a variety of exogenous influences, resulting in a continual increase in population rather than levelling off. Altogether, I am in the opinion that 'ye olde' logistic model, as an endogenous process model, did provide a basis for understanding the US population growth, but that discrepancies between the model and the observed, as brought out in Figure 4.4c, effectively guided us to find possible exogenous sources of the disagreements.

The lesson here is that the real value of a good theoretical model is not how well it fits an observed pattern. Paradoxically, its usefulness depends on how much insight we may gain from it when, actually, it did not agree with the observation. With a good model, we gain insight by looking into the disagreement and trying to comprehend how and why it did not fit. It follows that a good model must be built upon ecological first principles, i.e. undoubtedly true or axiomatic attributes, e.g. the existence of an upper limit in a reproductive rate and within-population competition.

4.13 Application of an Endogenous–Exogenous Process Model to Wildlife Management

A model that incorporates exogenous factors, like (4.9), may also be applied to the management of endangered species. How can once-flourishing species go extinct like the legendary passenger pigeon, great auk, and Tasmanian tiger, to name just a few? An endogenous model would not provide much insight. We must incorporate the effect of exogenous processes. Generally speaking, an endogenous process model has a few parameters (constants). Then, the rule of thumb is to incorporate the exogenous effect into each one of these constants, as in model (4.9), or at least into the one that you are most concerned about. Caution should be taken, however: too much generalization might get you nowhere, as it may become mathematically intractable.

Unlike an endogenous process, there is no standard or systematic way to build a model of an exogenous process because of its haphazard nature. But by looking into discrepancies between an appropriate endogenous process model and actual observations, you may be able to assess the effect (or source) of exogenous influences. For example, if you have made a sufficiently long series of observations, you may be able to trace the cause by estimating the series $\{\varepsilon_t\}$ and looking for an environmental factor that is correlated with it. Conversely, if the species is somehow recovering, you would see the estimated values of $(R_m + \varepsilon_t)$ increase with t. Then, the time-plot of the series $\{X_t\}$ could be sigmoidal rather than a monotonic increase as expected in the (endogenous) logistic process. That is, even the non-negative series $\{x_t\}$ of a recovering population may, after a log-transformation, turns out to be sigmoidal.

90 · Reproduction Curves and Their Utilities

You may create an idealized exogenous model process and see by simulation how it influences the pattern of the resultant population series, etc. You would learn a lot from such exercises. Have fun! Meanwhile, I have one more issue to talk about to conclude the present chapter: the origins of the myths of the logistic law.

4.14 The Origins of the Myths of the Logistic Law

As I have shown earlier, the sigmoidality of the classical logistic curve has no ecological significance. But why has this iconic symbol of the classic persisted until now? I think it is most likely to have originated from the 1920 Pearl–Reed paper on the US census data. In particular, the way the authors conceived the sigmoidality appealed to their contemporary ecologists and has been accepted ever since. The detail is as follows.

When Pearl and Reed were working on the census data, the US population was still increasing in an (apparently) accelerated manner (Figure 4.4a). Thus, they reasoned *a priori* that: in general, a population would at first increase at a geometric rate, but the rate of increase should, at a certain point in time, slow down, and should eventually approach 0 asymptotically because of the limited supply of essential resources and space, hence resulting in a sigmoid growth curve. Then, they proposed an equation, their equation (ix), as a descriptive model that fitted their perception of sigmoidality nicely.

However, Pearl and Reed did not develop the equation on their own. Rather, they borrowed one from chemistry that had been developed to describe the so-called autocatalytic reaction, and thought that the (would-be) sigmoid growth of the US population should be analogous to the chemical process. Unfortunately, however, the analogy misled them, as I now argue.

To begin with, let me describe the chemical reaction process. [Note: The Pearl–Reed paper did not describe what the chemical reaction was like. So, I consulted a chemist friend of mine to get some idea.] What I understood is basically as follows.

Consider a mixed solution of two kinds of chemicals, say A and B, whose concentrations are [A] and [B]. In the solution, B acts as the catalyst to convert A to B, i.e. its own kind: hence the name 'autocatalysis'. To keep the argument simple, let me idealize that one molecule of A is converted to one molecule of B, such that the total number of molecules in the solution stays constant, say [K], throughout the reaction process: i.e. [A] + [B] = [K]. Consider further that the process starts (at time $t = 0$) with high [A] and low [B]. Then, [B] starts increasing slowly. As time goes on, [B] increases more rapidly as the reaction is a self-accelerating process. Meanwhile, [A] is accordingly depleted. So, at a certain point in time, the rate of increase in [B] begins to slow down and gradually and eventually [B] approaches [K] as [A] diminishes to zero. Hence, the plot of changes in [B] against time t produces a sigmoid curve, depending on the initial concentration $[B]_{t=0}$ relative to $[A]_{t=0}$.

This inspired Pearl and Reed to envisage that the US population grew initially like the Malthusian (geometric) progression but, at a certain point in time, began to deviate from the Malthusian because of the limited supply of resources. Thus, the authors attached a special emphasis on the inflection point of their model as the starting point of deviation.

Somehow, Pearl and Reed justified the analogy (with the chemical process) by remarking that: 'There is much that appeals to the reason in the hypothesis that growth of population is fundamentally a phenomenon like autocatalysis'. Undoubtedly, this perception at once appealed to their contemporary ecologists, and has since persisted for generations.

However, after having learned about the chemical reaction, I saw that the analogy was inappropriate. The major reason is that chemical A is the sole source of chemical B, and hence an increase in [B] with time depends solely on [A] at every point in time during the process which ends when A is completely converted to B, and that is it.

In contrast, a biological (including human) population would utilize whatever is available for reproduction, and the resources would in general be continually replenished. Yet, as the population increases, its (per-unit-time) rate of increase decreases to zero sooner or later because the process is controlled primarily by competition among individuals over the available (self-replenishing) resources, not by depleting the resources. That is, the analogy breaks down with respect to the crucial aspect of the process of population growth. Most seriously, the conventional notion of the 'carrying capacity of the environment' stems from inappropriately comparing the asymptotic equilibrium level of the logistic process to the limited and depletable supply of chemical [A].

4.15 Final Remarks of Chapter

Having read all this, you might say that the Pearl–Reed model would still serve well as a descriptive and forecasting purposes. In my opinion, however, the real value of a model depends on its applicability as an analytical tool. As I showed earlier, a good model would provide insight when, paradoxically, it did not fit the observed pattern. Conversely, if the model was not built on ecological principles, it would mislead us at some point during an application.

From this point of view, a critical look at my own theoretical model (3.10) reveals that it is rather limited in applicability: it is based on a particular assumption that the individuals of the population are spatially distributed independently of one another, i.e. Poisson-distributed. The model should be generalized, by relaxing the assumption, to make it more flexible for a wider range of applications. This I attempt to do in Chapter 5.

Appendix 4A: The Derivative of a Function Is a Measure of the Slope of the Curve Generated by the Function

Consider the function $f(u)$ of a generic variable u, and a curve it generates, i.e. the plot of $f(u)$ against u. Consider also that the curve is uninterrupted (continuous) and smooth (with no sharp bend) in the range of u we choose to consider. Figure 4A.1 illustrates an example. Now, recall the definition of a derivative in Appendix 3B in Chapter 3. To repeat:

$$df(u)/du = [f(u + \Delta u) - f(u)]/\Delta u \quad \text{in the limit } (\Delta u \to 0).$$

(3B.1 rpt)

Here, we consider the geometry of the above definition in Figure 4A.1.

First, select a point on the ascending section of the curve and mark it as P_0. Draw a slanted (solid) line that is just touching the curve at P_0 from above (or outside). Mark this line as L_0: that is, L_0 is the tangent line (i.e. the slope) of the curve at P_0. [Note: Because the curve is smoothly continuous, the tangent line L_0 is uniquely defined.] From P_0, draw a (dotted) line vertically downwards and mark the intersection with the u-axis as u_0. In addition, draw a horizontal (dotted) line through P_0. Its leftward extension intersects the $f(u)$-axis at the point which reads $f(u_0)$.

Second, draw a (dashed) line (say L_1) through P_0 but at a slightly lower angle than L_0 such that its upper section intersects the curve at the point P_1: the geometers call L_1 a 'secant' line. Draw a (dotted) vertical line downwards from P_1 and mark the intersection as u_1. Also, extend this vertical line upwards to intersect L_0, and mark this point of intersection as P_1'.

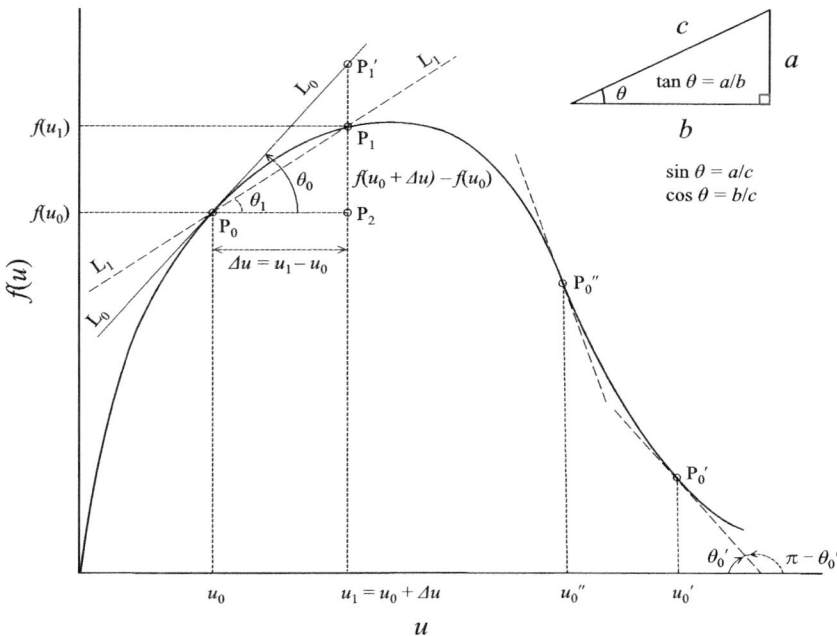

Figure 4A.1 Graphical explanation of the derivative defined in (3B.1 rpt).

Now, draw a horizontal (dotted) line from P_1 leftward and mark the reading of the $f(u)$-axis as $f(u_1)$. Further, extend this horizontal line through P_0 rightward and mark the intersection with the vertical line from P_1 as P_2. We see two right-angled triangles: triangle ($P_0 P_2 P_1'$) and triangle ($P_0 P_2 P_1$). Let θ_0 and θ_1 be the angles at the (common) vertex P_0 of the first and second triangles, respectively.

Further, let Δu be the distance between P_0 and P_2. Its length is equal to the difference between u_0 and u_1 on the u-axis, i.e. $\Delta u = u_1 - u_0$, or $u_1 = u_0 + \Delta u$. Then, we find that the distance between P_1 and P_2 is equal to $f(u_1) - f(u_0) = f(u_0 + \Delta u) - f(u_0)$.

Now, recall your school trigonometry, and apply it to the right-angled triangle as shown at the upper right corner of Figure 4A.1. The quotient a/b is a measure of the slope of the side c of the triangle. Recall, in particular, the trigonometric function $\tan \theta = a/b$. Then, applying it to triangle ($P_0 P_2 P_1$), we find:

$$\tan \theta_1 = [f(u_0 + \Delta u) - f(u_0)]/\Delta u. \tag{4A.1}$$

We are now going to have fun.

Using your imagination, move the point P_1 down towards P_0 along the curve. The vertical line through P_1 and P_2 moves with it leftwards as well. That is, u_1 moves towards u_0 on the u-axis, i.e. Δu tends to 0. As you do this, you see the secant line L_1 tending closer and closer (and eventually converging) to the tangent line L_0, and so does θ_1 to θ_0. In other words, in the limit $\Delta u \to 0$, the left-hand side of (4A.1) converges to $\tan \theta_0$ and so does the right-hand side to $df(u)/du$ by the afore-repeated (3B.1 rpt) at $u = u_0$. That is, in the limit ($\Delta u \to 0$) and at $u = u_0$, we find:

$$df(u)/du = [f(u_0 + \Delta u) - f(u_0)]/\Delta u = \tan \theta_0. \qquad (4A.2)$$

So, we understand that the derivative is a measure of the slope of the tangent line L_0.

In the above geometry, I used an example that is on the ascending section of the curve. What about a descending section? Consider a point on the rightwards tail of the curve, say P_0', and the angle of the tangent line, say $(\pi - \theta_0')$. If you follow the foregoing procedures, you would get the same results as in (4A.2): you just swap (u_0, θ_0) with $(u_0', \pi - \theta_0')$. But, then, the curve is decreasing at the point P_0' such that $f(u_0' + \Delta u) < f(u_0')$ or the term $[f(u_0' + \Delta u) - f(u_0')]$ in (4A.2) is negative. That is, the derivative is negative when the curve is decreasing. [Note: This is equivalent to the fact that the tangent line at P_0' has the slope whose angle (measured anticlockwise) is greater than 90°. Then, by the trigonometric identity, $\tan (\pi - \theta_0') = -\tan \theta_0'$, we see that the slope is negative.]

One more issue to note: the curve changes its curvature at the point P_0'': on the left of P_0'' it was convex upwards (or bulging up), whereas the curvature became concave up (or bulging down) on the right of P_0''. [Note: Convex up (or down) is the same as saying concave down (or up).] The point P_0'' is called an 'inflection point' as the curvature changes from convex up to convex down, or vice versa. How can we determine the position of this inflection point? Answer: by the position of the curve where its second derivative (i.e. the derivative of the first derivative) is equal to 0. The reason is the following. [Note: Incidentally, the second derivative is given by $d[df(u)/du]/du$. Mathematicians write this as $d^2f(u)/du^2$, or simply $f''(u)$.]

In Figure 4A.1, the curve $f(u)$ starts increasing with the steepest slope such that the corresponding derivative $f'(u)$ exhibits the highest positive value. As $f(u)$ increases to its apex, the slope of the curve becomes flat, i.e. $f'(u)$ tends to 0. As $f(u)$ begins to decrease past the apex, $f'(u)$ becomes negative and continues to decrease until reaching its lowest value at the inflection point P_0'', and then immediately begins to increase. In other words, if you plot $f'(u)$ against u, the $f'(u)$ curve reaches its trough at the inflection point of the original $f(u)$ curve. Now, remember that the slope of a curve at its trough is flat, meaning that its derivative = 0. That is, the derivative of the $f'(u)$ curve, i.e. the second derivative $f''(u)$, is 0 at the inflection point.

Appendix 4B: The Derivatives of a Few Standard Functions

I first show a few basic rules that would be handy for you. If you memorize them, you do not need to do calculations from scratch. You can come back here to remember them. Consider that the function $f(u)$ may comprise two different functions f_1 and f_2, both being differentiable (or 'well-behaved' in mathematicians' jargon), meaning that each of their curves are smoothly continuous, i.e. there is no break or sharp point, in the interval with which you are concerned. I drop the expression (u) to save space.

Rules:

(a)	Sum of functions	$f = f_1 + f_2$:	$f' = f_1' + f_2'$.
(b)	Product of functions	$f = f_1 \times f_2$:	$f' = f_1' \times f_2 + f_2' \times f_1$.
(c)	Reciprocal of a function	$f = 1/f_1$:	$f' = -f_1'/f_1^2$.
	Variant of (b) and (c)	$f = f_1/f_2$:	$f' = f_1'/f_2 - f_1 \times (f_2'/f_2^2)$
(d)	Exponential function	$f = \exp(f_1)$:	$f' = f_1' \times \exp(f_1)$
(e)	Logarithmic function	$f = \ln(f_1)$:	$f' = f_1'/f_1$.

Derivatives of some standard functions:

(i)	$f = c$ (constant):	$f' = 0$.	[Note: a constant is differentiable forever, i.e. 0 after 0 after...]
(ii)	$f = u^n$:	$f' = n \times u^{n-1}$.	
(iii)	$f = 1/u^n$:	$f' = -n/u^{n+1}$.	This is a variant of (ii) in that $1/u^n \equiv n^{-n}$ and applying (ii), you get (iii).
(iv)	$f = c^u$:	$f' = c^u \ln(c)$.	Special case of $c = e$: $f' = e^u$ $\because \ln(e) = 1$ by definition.
(v)	$f = \ln(u)$:	$f' = 1/u$.	
(vi)	$f = \sin(u)$:	$f' = \cos(u)$	
(vii)	$f = \cos(u)$:	$f' = -\sin(u)$	

Exercises:

Q_1: What if $f = cu$? A: $f' = c$. This is because, setting $c = f_2$ and $u = f_1$, rule (b) combined with (i) and (ii) yields the result.

Q_2: What if $f = \tan(u)$? A: $1/\cos^2(u)$. Applying the variant of (b) and (c) with $f_1 = \sin(u)$ and $f_2 = \cos(u)$, and using the trigonometric theorem that $\sin^2(u) + \cos^2(u) = 1$, you get the result.

Application: how do you differentiate model (4.1) in the main text, i.e. $x_{t+1} = x_t \, r_m \exp(-cx_t)$, with respect to x_t?

Let $x_{t+1} = f$, $r_m \, x_t = f_1$, and $\exp(-cx_t) = f_2$. Then, applying the result in Q_1 in the above exercises, we find $f_1' = r_m$. As well, applying the rule (d), we find $f_2' = -c \exp(-cx_t)$. Then, applying the rule (b), we have the result (4.2).

Appendix 4C: L'Hôpital's Rule

Consider that each of $f(u)$ and $g(u)$ tends to $\pm\infty$ or 0 in the limit $u \to a$ such that $f(u)/g(u) = 0/0$ or $\pm\infty/\pm\infty$, i.e. indeterminate. Consider further that $f(u)$ and $g(u)$ are both differentiable at $u = a$, i.e. they have derivatives $f'(a)$ and $g'(a)$. Then, the rule asserts that $f(u)/g(u) = f'(u)/g'(u)$ in the limit ($u \to a$). The following combinations are possible:

(i) If $g'(a)$ was constant (non-zero and finite), the quotient $f'(a)/g'(a)$ depends on $f'(a) = 0$, or $\pm\infty$, or constant. That is, $f'(a)/g'(a) = 0$, $\pm\infty$, or a constant, respectively.

(ii) If $g'(a) = 0$ and $f'(a) = \pm\infty$ or constant, $f'(a)/g'(a) = \pm\infty$. If $f'(a) = 0$, $f'(a)/g'(a) = 0/0$, i.e. indeterminate again.

(iii) If $g'(a) = \pm\infty$ and $f'(a) \neq \pm\infty$, $f'(a)/g'(a) = 0$. If $f'(a) = \pm\infty$, $f'(a)/g'(a) = \pm\infty/\infty$, i.e. indeterminate again.

(iv) In case $f'(a)/g'(a)$ is indeterminate, repeat the procedure to evaluate $f''(a)/g''(a)$ and calculate $f''(a)/g''(a)$, and so on, until you find a determinable result.

Appendix 4D: Prototype Curve and Its Translation

Equation (4.8) in the main text can be rewritten by moving (transferring) R_m from the right to the left of the equality sign, i.e.

$$(R_t - R_m) = -\exp(C + X_t) \qquad (4D.1)$$

in which we can write $(R_t - R_m) \equiv V_t$ and $(C + X_t) \equiv U_t$, i.e.

$$V_t = -\exp(U_t).$$

This is the prototype curve on the (V_t, U_t) coordinate system in that its slope $dV_t/dU_t = -\exp(U_t)$ at any given point on the curve does not depend on R_m and C.

Now, consider that the R_m value is increased (or decreased), while keeping the C value unchanged. This means that the V_t-coordinate is brought down (or up): alternatively, it means that the prototype curve is brought straight upwards (or downwards) relative to the V_t-axis. In other words, a change in R_m means a vertical shift (translation) of the prototype curve. Similarly, a change in C value means a lateral translation of the prototype curve along the U_t-axis: an increased (decreased) C means a leftwards (rightwards) shift. Altogether, a change in the R_m or C value merely changes the position of the curve in the coordinate plane without changing the prototype shape.

5 · Generalization of the Logistic Model

5.1 Preamble

In Chapter 3, I developed the theoretical model (3.10) as a discrete-time (single-species) process and compared it with the classical (continuous-time) logistic model. To repeat it:

$$r_t = r_m \exp[-s(1-k)x_t] \qquad (3.10 \text{ rpt})$$

where r_t is the net per-capita reproductive rate defined as x_{t+1}/x_t in which x_t represents population size (e.g. density) at time (generation) t; r_m is the maximum (potential) reproductive rate of an individual when it has no competitor in the area s around it (Figure 3.1); and $0 < k < 1$ is a measure of the intensity of competition with another individual: the lower the value of k, the higher the intensity.

As has also been shown, the advantage of theoretical model (3.10) over the classical model is twofold. First, it is constructed in such a way that every parameter represents an ecological attribute of the population process. Second, it generates a variety of patterns in the temporal changes in population about the equilibrium level after a logarithmic transformation, i.e. $X^* = \ln(x^*)$, whereas the classical model produces only one: monotonic convergence to the level K. Nonetheless, (3.10) is still much limited in applicability. As already explained in Chapter 4, the dynamical (time-dependent) pattern that model (3.10) generates (cf. Figure 4.6) is controlled only by one parameter, namely the potential reproductive rate, $R_m = \ln(r_m)$. In the present chapter, I attempt to generalize model (3.10) to make it more flexible for a wider range of application.

5.1.1 Initial Motivation for Generalization

In the early 1980s, I was looking into the classical work of Professor Utida (pronounced Uh-chi-dah) of Kyoto University on his laboratory populations of the azuki bean weevil (*Callosobruchus chinensis*):

5.1 Preamble

I introduced this work of Utida's in my previous book (Royama, 1992). His experiments produced two aspects of the population process: one was reproduction curves, plots of r_t against x_t (figure 7.2 of Royama, 1992) and the other was the series of population changes, i.e. the time-plots of x_t against t (figure 7.12 of Royama, 1992). I tried to fit my (then newly developed) model (3.10) to these results and found that it fitted the reproduction curves quite well, whereas it failed to produce the observed pattern of the series by simulation.

In the meantime, I remembered the work of Michael P. Hassell (1975) of Imperial College, London, in which he proposed a model of single-species population processes, his equation (11). The model is of the form:

$$r_t = x_{t+1}/x_t = r_m/(1 + ax_t)^b$$

in which the r and x are my notations used in (3.10) but a and b are Hassell's original constants. With this model, I tried to generate a few series by simulation and found that one of them reproduced the pattern of the observed weevil series nearly perfectly (figure 7.13 in Royama, 1992).

Although the Hassell model in its mathematical form worked very well as a descriptive model, I could not find the ecological meaning of the constants a and b before I finished the writing of my 1992 book. So, in the book, I only suggested that the model could be derived by changing the spatial distribution of the population concerned from the Poisson, on which (3.10) was based, to some non-Poisson distributions. Soon after the book came out, I received a letter from Peter Rothery (a statistician, then with the British Antarctic Survey). Peter suggested that the probability distribution function, i.e. Pr(i) of (3.6) in Chapter 3 of the present book, be changed from the Poisson to a more general one, called the negative binomial distribution. Indeed, his result was of exactly the same form as Hassell's model. So, I thought the problem resolved.

After a while, though, I noticed a problem. To see it, let's go back to the basic formula (3.4) in Chapter 3. To repeat:

$$r_t = r_{(0)}\Pr(0) + r_{(1)}\Pr(1) + r_{(2)}\Pr(2) + \ldots + r_{(i)}\Pr(i) + \ldots$$

(3.4 rpt)

The probability function Pr(i) here was defined primarily as the distribution of individuals (competitors) within a circle around each individual (cf. Figure 3.1). If the individuals are distributed completely at random (i.e. Poisson-distributed) over their effective habitat space, then the competitors in a circle around a given individual are also Poisson-distributed with the same parameter value, the average density of the population over the habitat space. However, this would not generally hold if Pr(i) was non-Poisson-distributed: the distribution of competitors within a circle in general differs in parameter values from the distribution over the habitat space. In other words, given the population size at generation t across the habitat space (i.e. x_t), the size x_{t+1} in the following generation cannot be estimated by model (3.4).

So, I went back to the drawing board but then left it unresolved. After a while, in fact a few decades later, I gave it a second look and this time had a good result. In the following, I show how I did it. To begin with, let me show you what the negative binomial distribution is like.

5.2 Negative Binomial Distribution

In the Poisson distribution, the occurrence of each individual in the habitat space is completely independent of the locations of other individuals, i.e. purely at random, like raindrops on flat ground in calm weather. This is a rather particular type of distribution. In actual situations, animals are more likely aggregated in various ways (e.g. clumped, patchy, localized, and whatnot) even within their effective habitat. [Note: By 'effective habitat' I mean a space excluding non-utilizable spaces. In spruce budworm, for example, the foliage of firs (*Abies*) and spruces (*Picea*) is its effective habitat, as will be shown in Chapter 10 in the present book.] The negative binomial distribution is so versatile that it can describe varieties of aggregated patterns, even including the Poisson distribution as a particular case, which will be shown in due course.

To introduce this versatile device to you, let me first talk about the plain 'binomial distribution'. In most textbooks in statistics, this distribution is illustrated with a flip-a-coin (or a roll-a-die) trial, a.k.a. a binomial trial. But, here, let me use a different example to suit my purpose.

5.2.1 Binomial Trial

Consider that you go to a trapshooting range to have a round of t shots: in fact, you have a number of rounds. [Note: Trapshooting is a game of shotgun shooting of clay pigeons (miniature Frisbees made of clay) launched in front of you and flying away from you in random directions within a certain range of vertical and horizontal angles.] In each round of t shots, you might hit or miss all t shots or, in general, something in between: you hit the targets h times and miss m times, but always $h + m = t$ because you have only t shotshells in your pocket in each round. So, $h = t - m$ (or $m = t - h$) can be anything from 0 to t. Suppose also that your shooting skill is as much as p in terms of the probability of a hit, and let the probability of a miss be q. Thus, quite naturally, $p + q = 1$ because you either hit or miss and nothing else; forget about misfires. Keep in mind, though, that your shooting skill level is assumed to remain unchanged throughout the session. So, p (and hence q) stays constant. [How do we determine p? Consider the quotient h/t. Because h is a random variable, the realized value of the quotient varies from round to round, even if your shooting skill stays constant. Nonetheless, as you shoot an unlimited number of rounds, i.e. t is unlimitedly large (provided that you have a self-reproducing wallet), the quotient would be averaged out (or expected) to converge to a certain value (no larger than 1 because $h \leq t$). This certain (expected) value is p. Formally written: $p = E(h/t)$ where the symbol E stands for the 'expected value' or simply reads: 'expectation'.]

Now, we look into the probability of occurrence of every possible combination of hits and misses, given t. If you have only a few rounds of t shots, not every possible combination may be realized. However, if you repeated so many rounds to keep the statistics of the frequency at which a given combination occurred, its proportion (the total frequency of the combination divided by the total number of rounds) would tend to converge to its expected value, like the foregoing case for p. If we assume ideal conditions, e.g. your gun functioning perfectly, your skill level being kept constant, no disturbance by wind, no fatigue, no drinking, we can calculate such expectations without actual trials.

For example, consider that all t shots are hits (as you wish). Then, the probability of this event to occur is p^t. This is because, by the multiplication rule for probabilities (of independent events), it is p multiplied by itself t times, i.e. p^t. [Note: Evidently, the occurrence of a hit in each shooting does not depend on what has happened (hit or miss) in the previous shots, unless you lose your composure after missing too many targets; let's assume you don't. In other words, each hit (or miss) in the sequence of shots is an independent event.] So, consider that you are an excellent 95% shooter, i.e. $p = 0.95$, for a hobby shooter's standard. Then, for a round of 10 shots (i.e. $t = 10$), $p^t = 0.95^{10} = 0.5987\ldots \approx 0.6$. This means that the perfect score of 10 hits is expected to happen 6 times out of 10 rounds.

Now, consider that you missed once in a given round, i.e. $(t-1)$ hits and a miss. The probability of this to occur in each round is given by $p^{t-1}q$. However, there are t different ways of this incidence (event) to occur in a given round: the miss may occur at first, second, third, ..., or at the last t-th shot. That is, there are t possible ways for the combination of $(t-1)$ hits and a miss to occur. Thus, we find that, by the addition rule for probabilities of mutually exclusive events, the probability of any one of these possible combinations to occur is given by $t \times p^{t-1}q$. [Note: The event of the (only) miss to occur at a given point in the shooting sequence is said to be 'exclusive' in that, once it has occurred, it would not be expected to occur for the second time in the round. Thus, the probability of any one of the aforementioned possibilities to occur is the sum of the t realizations of $p^{t-1}q$, i.e. $t \times p^{t-1}q$.] In particular, for $p = 0.95$ (hence $q = 0.05$) and $t = 10$, we find $t \times p^{t-1}q \approx 0.315$. That is, the incidence of 9 hits and a miss is expected to happen about 3 times out of 10 rounds, or about 32 times out of 100 rounds.

In general, the probability of h hits and m misses to occur is given by $p^h q^m$: this is also equal to $p^{t-m}q^m = p^h q^{t-h}$, and you can use whichever expression is appropriate in the context of your arguments. Furthermore, the combination of h hits and m misses can occur in $t!/h!m!$ different ways. [Note: The symbol '!', called 'factorial' (not 'surprise!'), is defined for a non-negative integer, say t, such that: $t! = t \times (t-1) \times (t-2) \times (t-3) \times \ldots \times 2 \times 1$. As a special case, 0! (if h or $m = 0$) is defined to be equal to 1, i.e. $0! = 1$. Why? See Appendix 5A.]

Altogether, under the constraint $m + h = t$, the probability of a given combination of hits and misses to occur is neatly given by each term of the binomial series as the expansion of $(p + q)^t = 1$, i.e.

$$(p+q)^t = p^t + tp^{t-1}q + \ldots + (t!/h!m!)p^h q^m + \ldots + tpq^{t-1} + q^t = 1.$$
(5.1)

Hence the name 'binomial distribution'. Let us do one more calculation for the probability of the combination of 8 hits and 2 misses to occur, while keeping the probability of hit $p = 0.95$ as before. Substituting $h = 8$, $m = 2$, $p = 0.95$ and $q = 0.05$ in the general term in binomial expansion (5.1), we see that $p^8 q^2 = 0.95^8 \times 0.05^2 \approx 0.001659$ and $10!/8! \times 2! = 10 \times 9/2 = 45$. Hence, the probability is about $0.001659 \times 45 \approx 0.075$. That is, we expect to see this happen in 7 or 8 rounds out of 100 rounds of 10 shots.]

5.2 Negative Binomial Distribution · 103

The coefficient $(t!/h!m!)$ of the general term in (5.1) is called the 'binomial coefficient', and the general term $(t!/h!m!)p^h q^m$ is called a 'probability mass function' (of the binomial trial) which is the 'probability distribution function' of the discrete random event h or m with the corresponding probability p or q. [Note: The term 'probability *mass* function' of discrete random events is the counterpart of the 'probability *density* function' of continuous random measurements, e.g. the well-known normal distribution of, say, body weights among the (mono-gender) students in a class.] The mass function gives the probability of every possible outcome of the game, its frequency of occurrence being distributed over the entire range of h or $m = 0, 1, 2, \ldots t$, given $m + h = t$. Therefore, all terms on the right-hand side of (5.1) add up to 1.

So much for the binomial distribution. Let's move on to the main theme, the negative binomial distribution. Our goal is to find the probability mass function of the negative binomial, i.e. the counterpart of the general term $(t!/h!m!)p^h q^m$ of the binomial distribution (5.1).

5.2.2 Negative Binomial Trial

To this end, let us change the rules of the game. Rather than using up all t shotshells in your pocket to complete one round of shooting, we set a goal of hitting first h targets per round of shooting. That is, you keep shooting until the goal is achieved to complete one round. If you are a good shooter, you would achieve it quickly. But, if you are a poor shot like me, you may have to keep shooting. Therefore, while the number h is fixed as a goal to achieve, the number m of missing targets can be anything from zero to whatever you have to keep shooting until you hit h targets to finish one round of the game. In theory, the goal may never be achieved; even if you kept shooting from dawn to dusk, a truck load of shotshells may not be enough. In other words, the number of misses m may vary from 0 to ∞ (the symbol reads: 'infinity'), depending on your skill and luck. We now look into the probability of achieving the goal under the assumption, as before, that your skill of hitting a target remains p in every session. Here is how to calculate it.

Calculation of probability mass function. Obviously, you must have hit $(h - 1)$ targets before you achieve the goal of h hits. But, even having reached that stage, you could have missed many targets as well. Besides, even after you have hit $(h - 1)$ targets, and are anxiously hoping for hitting the final h-th target, you may generally miss many more targets. In other words, you must have shot $(h - 1)$ hits and m misses just

104 · **Generalization of the Logistic Model**

prior to achieving the goal with one more hit. Thus, the total number of shots up to this stage is the sum $h - 1 + m$, with m varying from 0 to infinity. Then, the probability of reaching this stage is given by $p^{h-1}q^m$ multiplied by the number of all possible combinations of the occurrences of $(h - 1)$ hits and m misses.

You can calculate the number of combinations by analogy of the binomial coefficient, $t!/h!m!$, given by the general term in (5.1), but with the following modifications: (i) the total number of shots t is now replaced by the random number $(h - 1 + m)$, i.e. no longer fixed but varying as m varies; (ii) the total number of hits this far (i.e. one more hit remains) is now equal to $(h - 1)$, i.e. a constant.

Replacing the expressions t ($= h + m$) and h, in the binomial coefficient $(t!/h!m!)$ in (5.1), with the expressions $(h - 1 + m)$ and $(h - 1)$, respectively, we find that the number of possible combinations of $(h - 1)$ hits and m misses out of the total $(h - 1 + m)$ shots is given by:

$$(h - 1 + m)!/(h - 1)!m!.$$

Thus, the probability of the occurrences of $(h - 1)$ hits and m misses is given by $p^{h-1}q^m$ multiplied by $(h - 1 + m)!/(h - 1)!m!$, i.e.

$$[(h - 1 + m)!/(h - 1)!m!]p^{h-1}q^m.$$

Then, imagine that your dream finally comes true with the last $(h + m)$-th shot that is a hit with the probability p. Thus, the probability of achieving the goal that we have been looking for is given by the above probability multiplied by one more p, i.e.

$$\{[(h - 1 + m)!/(h - 1)!m!]p^{h-1}q^m\} \times p = [(h - 1 + m)!/(h - 1)!m!]p^h q^m.$$

This probability varies depending only on m, because the other parameters are all fixed. Hence, it can be written simply as $\Pr(m)$. Thus, the probability of achieving the goal after missing m shots is given by:

$$\Pr(m) = [(h - 1 + m)!/(h - 1)!m!]p^h q^m, \quad (m = 0, 1, 2, \ldots, \infty). \quad (5.2)$$

This is the probability mass function of the negative binomial distribution, giving the probability of every possible outcome of the game that is distributed over $m = 0, 1, 2, \ldots, \infty$. [Note: The notations h, m, and $q = 1 - p$ are chosen to facilitate my argument here. These differ from those you would see in standard textbooks of statistics and probability theory: my (h, m, q) are usually written as (r, k, p). However, in the present book, I use r and k to denote, respectively, the rate of increase (generally changes) in population and the intensity of competition. So, I use the different notations.]

Because the Pr(m) is a probability mass function, the right-hand side of (5.2) sums up to 1 over $m = 0$ to infinity, remembering that the probability of all possible events to occur is 1. Thus,

$$p^h \Sigma[(h - 1 + m)!/(h - 1)!m!]q^m = 1, \quad (m = 0, 1, 2, \ldots, \infty). \quad (5.3)$$

[Note: The term p^h, which is placed at the end of formula in (5.2), is moved to the head in (5.3), i.e. outside (left of) the summation sign Σ, because it is a constant, independent of m. Also, why the name 'negative' binomial? If you are curious, see Appendix 5B.]

Some statistical properties of negative binomial distribution. Now that the number of misses, m, is a random variable, its distribution is characterized by its moments, in particular by the first two, i.e. mean (say, μ, the lower-case Greek mu) and variance (σ^2, the familiar lower-case Greek sigma). These are determined in terms of the parameters (constants) p and h. In other words, μ and σ^2 can be expressed in terms of p and h. The results are shown to be:

$$\mu = h(1 - p)/p \text{ and } \sigma^2 = h(1 - p)/p^2. \quad (5.4)$$

[Appendix 5C shows the derivation of these relationships.] Conversely, solving the first of the above relationships for p and substituting the result for the second, we have:

$$p = h/(\mu + h) \text{ and } \sigma^2 = \mu(1 + \mu/h). \quad (5.5)$$

I will use these results in due course.

5.3 Ecological Application of the Negative Binomial Distribution

At this point, you might ask: How is this probability distribution in trapshooting pertinent to ecology? [Answer: I am talking about competition. Well, stop that cheap shot.] As it turns out, the negative binomial distribution very well fits varieties of the spatial distribution of organisms, e.g. trees of a given species in a forest stand. It is a very flexible model and describes varieties of aggregated (patchy, clumped, clustered, and whatnot) distributions, even including the purely random (unaggregated) Poisson distribution as a particular (limiting) case in which h tends to infinity, as will be shown later. In fact, silviculturists have been using the negative binomial model for estimating densities, patchiness, etc. of trees in a stand quite successfully.

You might still wonder, though, why I went through all the trouble with the trapshooting competition to derive the probability mass function. If the final aim was the derivation of a model of spatial distributions, why should I not have gone at it directly? The reason is that, as applied to a spatial distribution, the negative binomial distribution is an empirical (or descriptive) device, as it were. Because it fits so many different types (in their ecological generating mechanisms) of aggregated distributions, it would not be easy to formulate it on a general, unspecified ground. Therefore, a conventional way of introducing it with a concrete example (like trapshooting or more usually coin tossing) provides a good starting point before considering its empirical applications. An important lesson here is that an empirical model, borrowed from something unrelated, is often (if not always) found to be useful in ecological studies. It is sort of a reversal of the lesson we learnt in Chapter 4, i.e. the analogy of logistic process with autocatalysis: the similarity of models in mathematical form does not necessarily imply the same underlying mechanism. So, we should be flexible in choosing between the two approaches.

Now, to apply the negative binomial distribution in trapshooting to the spatial distribution of organisms, we need to translate the parameters involved in the probability mass function (5.2) into those of the spatial distribution.

5.3.1 Translation of Trapshooting Parameters into those of the Spatial Distribution

As already explained, the individual-centred scheme of distribution in Figure 3.1 is no longer appropriate in dealing with the negative binomial (a non-Poisson) distribution. So, I suggest instead that we employ the following quadrat scheme.

Suppose that we divide a forest stand (a sheet of graph paper will do) into many (non-overlapping) quadrats of an equal size and count the number of individual trees of a given species (as dots) in each quadrat. Then, m in (5.2) corresponds to the counts of individual trees in a quadrat, with the mean μ and the variance σ^2 in (5.4). Thus, the probability $\Pr(m)$ of (5.2) in the shooting game corresponds to the expected proportion of the quadrats, each containing m trees. It sounds rather strange, you might say, to compare the number of missed targets to the number of trees actually present in the quadrat. But that is a mathematical equivalence in that a count is a count: the model would not care what you are actually counting; let's not argue. Nonetheless, to avoid confusion, I will change the letter m to j, and the designation $\Pr(m)$ of trapshooting, to $\text{Qr}(j)$ in the quadrat model. That is, the expected proportion (probability) of the quadrats, each containing j trees, is $\text{Qr}(j)$.

But what does the parameter h in (5.2), the number of hits as the goal of the shooting game, correspond to in the quadrat system? It is translated into an index for the degree of aggregation of the trees for the following reason.

Generally in statistics, the quotient σ^2/μ (i.e. the *variance/mean* ratio), called the index (coefficient) of dispersion, is used as a measure of the degree of aggregation of things or events in space or in time. This is based on the fact that in the purely random, unaggregated Poisson distribution, the quotient is equal to 1 because, in this particular distribution, $\sigma^2 = \mu$ (usually designated by the Greek λ) is the only parameter characterizing this distribution. In an aggregated distribution, however, the variance is greater than the mean, i.e. $\sigma^2/\mu > 1$. In the present case, using the second relationship in (5.5), we find:

$$\sigma^2/\mu = 1 + \mu/h.$$

Clearly, the right-hand side is greater than 1 for a finite $h > 0$. In particular, given μ, the smaller the h-value, the more aggregated. Conversely, as h tends to infinity, the coefficient will converge to 1, i.e. converging to the unaggregated Poisson distribution. In other words, the Poisson distribution is a special (limiting) case of the negative binomial distribution in which h tends to infinity.

Now, there is an important point to note in the translation of trapshooting to spatial distribution. Although defined as a non-negative integer in trapshooting, h (usually designated r or k in the literature on statistics) is generalized such that it can assume a positive (non-integer) value. It is strange to think of 0.3 hits as the goal to achieve in trapshooting: not that you chip off a 0.3 part of the clay pigeon, but it is a mathematical abstraction to broaden the range of its application. After all, we can think of 0.3 hits as an average value. So, be flexible. [Digression: Mathematicians do weird things. But, think about it, it was not too long ago when a negative number was considered to be weird, let alone an imaginary number which, when squared, would yield a negative number: in fact, even many mathematicians of the time thought it was utter nonsense. Nonetheless, they never hesitate to accept

108 · Generalization of the Logistic Model

such a weird, unreal, or even ridiculous thing, once they find its utility in their abstract system of thinking. So, I follow their philosophy in the present book to justify a weird thing I would do in the next Chapter 6, for example, where I apply the negative binomial model that has been developed here.]

This completes the translation of the parameters in trapshooting into those in the spatial distribution of trees (or animals, of course).

5.4 A General Model of Intraspecific Competition

Usually in describing the spatial distribution of things, as in a silvicultural survey, the size of a quadrat would be determined arbitrarily for a technical convenience. It only needs to ensure an adequate degree of resolution to distinguish among different types of distribution: evidently, given the density of the objects concerned, too large or too small a quadrat may result in the inability to distinguish between different types.

In building a model of competition, however, a special consideration is required. I divide the effective habitat space into n quadrats of equal area s which, as defined in Figure 3.1, is (four times) the area that contains a minimum sufficient amount of resources for an individual to survive and reproduce normally. Thus, we assume that all individuals (dots) in a given quadrat (of area s) compete with each other. It is important to bear in mind that the location of a quadrat should be independent of where each dot is; if not, the very reason for avoiding the individual-centred scheme of Figure 3.1 would be lost. Generally, though, individual quadrats (as unit spaces for describing a spatial distribution) could be in any shape: circle, triangle, or whatever, as long as they are of the same size, non-overlapping, and their locations are independent of the distribution of the dots. However, a regular (squared) grid pattern, like the usual graph paper, would be the easiest one for visualizing the model distribution in mind.

5.4.1 Formulation of the Quadrat Model

Let x_t be the density of the individuals at generation t distributed over the habitat space divided into n quadrats. Then, the average number of individuals in each quadrat (of size s) is sx_t, or the total number in the habitat space is nsx_t. Likewise, the total number of offspring produced is nsx_{t+1}: no migrations to avoid complications without compromising

5.4 General Model of Intraspecific Competition · 109

generality. [Note: In the analysis of the variation of a population in time, its density is a variable, a function of t. When dealing with the spatial distribution of a population at a given point in time (e.g. a given developmental stadium within generation), its density is a parameter (fixed constant) that characterizes the distribution.]

Now, let $Qr(j)$ be the expected proportion of the quadrats, each being occupied by j (= 0, 1, 2, ...) individuals. [Note: unlike the model in Figure 3.1, no single individual is picked up; all are equal indiscriminately.] Consider further that these j individuals within the same quadrat compete with each other: that is, each individual in the quadrat has (j − 1) competitors. Furthermore, assume, as we did in the process of arriving at relationship (3.5) in Chapter 3, that the mean potential reproductive rate r_m (of an individual with no competitor) would be reduced by the positive factor $k < 1$ when competing with another individual. Hence, the reproductive rate of each individual in a quadrat containing j individuals, each competing with (j − 1) competitors, is expected to be $r_m k^{j-1}$. However, over the whole habitat space (sheet of graph paper), there are $Qr(j)n$ quadrats, each containing j individuals, and hence there are $jQr(j)n$ individuals altogether. Thus, the total number of offspring produced in all n quadrats (i.e. nsx_{t+1}) is given by the sum of the terms $nr_m jk^{j-1}Qr(j)$ over $j = 0, 1, 2, \ldots$, to infinity, i.e.

$$nsx_{t+1} = nr_m \big[Qr(1) + 2kQr(2) + 3k^2 Qr(3) + \ldots \\ + jk^{j-1}Qr(j) + \ldots\big]; \quad (j = 1, 2, \ldots, \infty). \quad (5.6)$$

[Note: Although the above series is infinitely long, the terms $jk^{j-1}Qr(j)$ would practically vanish for sufficiently large j. This is firstly because $Qr(j)$ would vanish for a large j: it is unlikely to happen that, given the average number of individuals sx_t, a quadrat is jam packed with an infinite number of animals or trees. Secondly, the quantity jk^{j-1} would also vanish because $0 < k < 1$ as defined: see Appendix 5D for details. In other words, the sum in the brackets on the right-hand side of (5.6) converges to a certain finite value.]

Thus, the average per-capita reproductive rate r_t from generation t to $t+1$ (i.e. $r_t = x_{t+1}/x_t$) in the population as a whole is given by dividing both sides of (5.6) by the total number of parent individuals (nsx_t), such that we have:

$$r_t = r_m \big[Qr(1) + 2kQr(2) + 3k^2 Qr(3) + \ldots + jk^{j-1}Qr(j) + \ldots\big]/sx_t. \quad (5.7)$$

This is the within-quadrat competition model to replace the individual-centred model (3.5). [Note: Model (3.5) remains valid as a particular case of (5.7) in which Qr(j) is Poisson-distributed.] We are now ready to formulate a general model of competition.

5.4.2 Formulation of Competition Model

The process of formulation is somewhat tedious. Maybe, it's going to be like a cross-country endurance run; maybe, a little harder than the trapshooting exercise. However, generally speaking, ecological model building often requires a tedious process of algebraic manipulations which we the population ecologists must get used to. So, let's go.

Consider that the individuals are negative-binomial distributed over the effective habitat space. Then, replacing m with j and Pr(m) with Qr(j) in model (5.2), we find the expected proportion of the quadrats that contain j (parent) individuals to be:

$$\text{Qr}(j) = p^h[(h-1+j)!/(h-1)!j!]q^j.$$

Substituting the right-hand side of the above relationship for the Qr(j) in (5.7), and noting that $j/j! = 1/(j-1)!$, we find after a little algebraic manipulation:

$$r_t = r_m p^h \Sigma j[(h-1+j)!/(h-1)!j!]k^{j-1}q^j/sx_t$$
$$= r_m p^h \Sigma\,[(h-1+j)!/(h-1)!(j-1)!]k^{j-1}q^j/sx_t;$$
$$\text{summation is over } j = 1, 2, \ldots, \infty. \quad (5.8)$$

Now, our goal here is to evaluate the right-hand side of (5.8) in terms of the parameters r_m, h, k, s, and x_t. That is to say, we want to somehow get rid of the random number j and the parameters $p = 1 - q$ on the right-hand side. Let us get rid of j to begin.

For this purpose, I rewrite the second right-hand side of (5.8) to make it a more convenient form. The expression q^j/sx_t at the very end of the left-hand side of (5.8) is manipulated to $(q/sx_t)q^{j-1}$ and the expression (q/sx_t) is moved to the left of (outside) the summation sign Σ, as it does not involve j, such that:

$$r_t = r_m\left(p^h q/sx_t\right)\Sigma[(h-1+j)!/(h-1)!(j-1)!]k^{j-1}q^{j-1}. \quad (5.9)$$

Now, given that sx_t is the average number of individuals per quadrat over the habitat space, it is equal to the mean of the random number j. In

5.4 General Model of Intraspecific Competition

other words, it is equivalent to the mean μ in (5.4). Thus, after replacing μ with sx_t, we have from (5.5): $p = h/(sx_t + h)$, $q = 1 - p = sx_t/(sx_t + h)$, and hence $q/sx_t = p/h$. Thus, the quotient in the parentheses outside (in front of) the summation sign in (5.9) becomes:

$$p^h q/sx_t = p^{h+1}/h.$$

Substituting (p^{h+1}/h) for $(p^h q/sx_t)$ in (5.9), and after the minor cosmetic change of the expression $(h - 1 + j)$ to $(h + j - 1)$, we have:

$$r_t = r_m(p^{h+1}/h)\{\Sigma[(h+j-1)!/(h-1)!(j-1)!](kq)^{j-1}\}.$$

Notice further that the expression $[1/(h-1)!]$ on the right of the Σ sign, multiplied by $(1/h)$ just outside the Σ sign, i.e. $1/[h(h-1)!]$, yields $(1/h!)$. Thus, the above relationship is reduced to:

$$r_t = r_m p^{h+1}\{\Sigma[(h+j-1)!/h!(j-1)!](kq)^{j-1}\},$$
$$\text{summed over } j = 1, 2, \ldots, \infty. \qquad (5.10)$$

Now comes a bit of surprise. As it turns out, the summation inside the braces on the right-hand side of (5.10) converges (in the limit $j \to \infty$) to $(1 - kq)^{-(h+1)}$, i.e.

$$\Sigma[(h+j-1)!/h!(j-1)!](kq)^{j-1} = (1-kq)^{-(h+1)}, \quad j = 1, 2, \ldots, \infty.$$

Appendix 5E gives the algebraic detail of the surprise: I hope you read the Appendix rather than just swallow the above result. Substituting $(1 - kq)^{-(h+1)}$ for the summation in (5.10), we have:

$$r_t = r_m p^{h+1}(1-kq)^{-(h+1)} = r_m[p/(1-kq)]^{(h+1)}. \qquad (5.11)$$

So, we have gotten rid of j. Next, let's get rid of $p = 1 - q$ on the right-hand side of (5.11).

To do this, recall the equality $p = h/(\mu + h)$ in (5.5), in which $\mu = sx_t$ as already mentioned. Hence, we find: $p = h/(sx_t + h)$ and $q = 1 - p = sx_t/(sx_t + h)$. Substituting these for p and q in the quotient in the brackets on the right-end of (5.11), and after a little algebra (your homework), we find:

$$p/(1-kq) = [h/(sx_t + h)]/\{1 - [ksx_t/(sx_t + h)]\}$$
$$= 1/[1 + (1-k)sx_t/h].$$

We see p and q no longer appear on the right-hand sides. Substituting the above result back in the right-end of (5.11), we find:

$$r_t = r_\mathrm{m}/[1 + (1-k)sx_t/h]^{h+1}, \qquad (5.12a)$$

or writing the expression $(1-k)s$ just as c as we did before, model (5.12a) can be compactly written:

$$r_t = r_\mathrm{m}/[1 + cx_t/h]^{h+1}. \qquad (5.12b)$$

Well, we finished the cross-country run: (5.12) is the general model of competition that we attempted to formulate. I now show that the discrete-time logistic model (3.10) developed in Chapter 3 is a particular case of (5.12).

5.5 Model (3.10) as a Particular Case of Model (5.12)

Recall that model (3.10) is based on the Poisson distribution. However, as already noted in the Section 5.3.1, the Poisson distribution is a limiting form of the negative binomial distribution (5.2) in which $h \to \infty$. In other words, we should have (3.10) by letting h in (5.12) tend to infinity. Let's see.

For convenience, let us write $cx_t \equiv w$ and $w/h \equiv 1/h'$, temporarily, such that $h \equiv h'w$. Substituting these for the corresponding terms in (5.12), and after a little manipulation, we can rewrite the denominator of the right-hand side of (5.12b) as:

$$[1 + 1/h']^{1+h'w} = [1 + 1/h']\left\{[1 + 1/h']^{h'}\right\}^w.$$

Now, let h (and hence h') tend to infinity. Then, the term $[1 + 1/h']$ on the right-hand side converges to 1, whereas $[1 + 1/h']^{h'}$ converges to the familiar but enigmatic number e, as already noted in the definition of e in (3B.4) in Appendix 3B of Chapter 3. Thus, in the limit ($h' \to \infty$), the expression on the right-hand side of the above equation becomes e^w. Now that $w \equiv cx_t$, we can write $e^w \equiv \exp(w) \equiv \exp(cx_t)$. Thus, substituting this result for the denominator on the right-hand side of the model (5.12), we find:

$$r_t = r_\mathrm{m}/\exp(cx_t)$$

and, further using the identity $1/u \equiv u^{-1}$, we make a little cosmetic change as:

$$r_t = r_\mathrm{m}\exp(-cx_t)$$

which is identical to model (3.10) in which $s(1 - k) \equiv c$. Thus, we understand that (5.12) is a general model of intraspecific competition in the single-species population processes. However, because the classical (continuous-time) logistic model is a particular case of the discrete-model (3.10), we see that model (5.12) is an even more general form (Mother) of the logistic model.

5.6 Interpretation of the Hassell Model: $r_t = x_{t+1}/x_t = r_m/(1 + ax_t)^b$

At first glance, it is clear that the Hassell model is identical in form to (5.12). Hence, by parameter-by-parameter comparisons, we can interpret a and b in terms of the corresponding parameters in model (5.12): that is, to translate (a, b) into (c, h) in (5.12). The results are:

$$a = c/h \equiv s(1 - k)/h \text{ and } b = (h + 1). \qquad (5.13)$$

Thus, I thought I had successfully interpreted the ecological significance of the Hassell model. But, after a while, I found a problem: the comparisons (5.13) suggest that a and b are not independent parameters, as both involve h. Why this is problematic is as follows. To investigate the nature of intraspecific competition with his model, Hassell considered a situation in which $0 < b < 1$, but this implies $0 < (h + 1) < 1$, and so the parameter h must be negative (a negative hit in trapshooting?). So, I had a dilemma. After a while, I realized that $0 < b < 1$ is ecologically feasible and should be accepted. It implies that a negative h should be accepted, too, albeit as weird as it may sound at this moment. In Chapter 6, in which I delve into the nature of competition, I generalize the ecological meaning of h so as to allow it to be negative. After all, I find this generalization is important as it opens the door to a hitherto unexplored area of ecology. However, before we move on to Chapter 6, there is one more model that requires our attention.

5.7 One More Model to Examine

More recently, the mathematicians, Å. Brännström and D. J. T. Sumpter, of Umeå University, Sweden, jointly published an article (Brännström and Sumpter, 2005) in which they gave a comprehensive review of mathematical models of competition. In the article, the authors derive

114 · Generalization of the Logistic Model

a model, their equation (3.4), which is identical in form to my (5.12): or rather, mine is identical to theirs, as theirs was published ahead of mine. The identicalness is due to the fact that both models are based on the negative binomial distribution. After careful examination, however, I found a difference between the two models in an ecological attribute which is in short as follows.

The Brännström–Sumpter model is given in the form:

$$a_{t+1} = b\lambda^{1+\lambda}a_t/(\lambda + a_t/n)^{1+\lambda} \tag{5.14}$$

where a_t represents the total number of individuals in generation t over n discrete (resource) sites (equivalent to my quadrats), and λ represents the (negative binomial) parameter that is equivalent to h in my (5.12). Now, dividing both numerator and denominator on the right-hand side of (5.14) by $\lambda^{1+\lambda}$, and then dividing both sides by a_t, model (5.14) can be rewritten as:

$$a_{t+1}/a_t = b/(1 + a_t/n\lambda)^{1+\lambda}$$

which is directly comparable in form with my (5.12a). Thus, term-by-term comparisons reveal that: $a_{t+1}/a_t = x_{t+1}/x_t = r_t$, $b = r_m$, $a_t/n = sx_t$, and $\lambda = h$. We see that their expression $a_t/n\lambda$ corresponds to my sx_t/h. In other words, their model is a particular case of (5.12b) in which $c = s$, rather than the full expression $s(1 - k)$ in (5.12a). This means that $k = 0$ identically in (5.14). However, $k = 0$ is an extreme case ecologically as shown below.

Setting $k = 0$ in (5.7), you would see all terms except Qr(1) inside the brackets on the right-hand side vanish. This means that only those individuals that occupy a quadrat singly would produce offspring, whereas none could reproduce in those quadrats occupied by a plural number of individuals; even if just two are present in a quadrat, both would completely fail to reproduce, which is an extreme idealization.

Notwithstanding, in (5.12), k and s form the single positive parameter $s(1 - k) \equiv c$, and the value of c could be estimated by observation, regardless of the value of k. In other words, the Brännström–Sumpter model creates no problem if used as a descriptive model. However, non-zero values of k would become a matter of importance if the model is to be used as an analytical device: that is to say, when the effect of competition is explicitly considered in an ecological investigation, as I attempt in Chapter 6.

Appendix 5A: Why 0! = 1?

Take a look at the general term of the binomial coefficient in (5.1) in the main text, i.e. $(t!/h!m!)$, and consider the particular case of $h + m = t = 3$ for example. There are four cases of combinations with $h + m = 3$: (i) 3 hits and 0 misses; (ii) 2 hits and 1 miss; (iii) 1 hit and 2 misses; and (iv) 3 misses and 0 hits. These cases occur in the following arrangements, with the number of combinations in each case being given by the coefficient $3!/(h! \times m!)$ for $h + m = 3$. Thus, we find all possible combinations and their corresponding binomial coefficients to be:

	Combinations	Binomial coefficients
(i)	(h h h); m = 0	$3!/(3! \times 0!) = 1$
(ii)	(h h m), (h m h), (m h h)	$3!/(2! \times 1!) = 3$
(iii)	(h m m), (m h m), (m m h)	$3!/(1! \times 2!) = 3$
(iv)	(m m m); h = 0	$3!/(0! \times 3!) = 1$

Evidently, for the binomial coefficients in (i) and (iv) to hold, we must define $0! = 1$.

Appendix 5B: Why the Name 'Negative Binomial'?

There are a few different ways of explaining it. The following is my unorthodox way of conceiving it. Recall that the usual binomial distribution is given as the expansion of $(p + q)^t = 1$ as in (5.1) in the main text, whereas the negative binomial distribution is given in the form (5.3). To repeat it:

$$p^h \Sigma [(n - 1 + m)!/(h - 1)!m!] q^m = 1. \qquad (5.3 \text{ rpt})$$

Noting that $p = (1 - q)$ on the left-hand side and dividing both sides by $p^h = (1 - q)^h$, we have:

$$\sum [(h - 1 + m)!/(h - 1)!m!] q^m = (1 - q)^{-h}.$$

We see then that the negative binomial distribution (5.3) can be viewed as the expansion of the expression $p^h (1 - q)^{-h} = 1$, which can be manipulated to:

$$[(1/p) + (-q/p)]^{-h} = 1.$$

116 · Generalization of the Logistic Model

Then, the mass function of the negative binomial distribution can be viewed as an expansion of the above formula. However, the formula is identical in form to $(p + q)^t = 1$ for the binomial distribution but the exponent $(-h)$ is negative. Hence the name.

Appendix 5C: How to Calculate the Mean and Variance of the Random Number m in (5.4)

(i) Calculation of the Mean $\mu = h(1 - p)/p$

Consider that we recorded the frequency of misses occurring 0, 1, 2, ... times, designated $\text{Fr}(m = 0, 1, 2, \ldots)$, say. Then, the sum of all realized values of m, weighted by the frequency occurrence of each value, is given by:

$$[0 \times \text{Fr}(0) + 1 \times \text{Fr}(1) + 2 \times \text{Fr}(2) + \ldots + m \times \text{Fr}(m) + \ldots]$$
$$= \Sigma m\text{Fr}(m), m = 0, 1, 2, \ldots$$

Thus, $\Sigma m\text{Fr}(m)$, divided by the total frequencies $\Sigma \text{Fr}(m)$, gives the average of m, i.e.

$$\text{Average of } m = \Sigma m\text{Fr}(m)/\Sigma\text{Fr}(m). \tag{5C.1}$$

If we repeat the trials a large enough number of times, the proportion $\text{Fr}(m)/\Sigma\text{Fr}(m)$ would converge to its expectation, i.e. the probability $\text{Pr}(m)$ given in (5.2) in the main text. Thus, the mean of m (i.e. μ) is given by replacing the $\text{Fr}(m)$ in (5C.1) with the $\text{Pr}(m)$ in (5.2). So, noting that $\Sigma\text{Fr}(m)$ converges to 1 as does $\Sigma\text{Pr}(m)$, and that $m/m! = 1/(m-1)!$, we find:

$$\mu = \Sigma m\text{Pr}(m) = \Sigma m[(h - 1 + m)!/(h - 1)!m!]p^h q^m$$
$$= \Sigma[(h - 1 + m)!/(h - 1)!(m - 1)!]p^h q^m. \tag{5C.2}$$

The key here is to somehow eliminate the random variable m on the right-hand side because we are looking for its average value, μ, in terms of h and $p = 1 - q$. The trick is to modify the expression $p^h q^m$ to $(p/q) \times p^{h+1}q^{m-1}$ and $1/(h-1)!$ to $h/h!$ such that (5C.2) is modified (without changing the mathematical relationship) to:

$$\mu = (hq/p)\Sigma[(h + m - 1)!/(h)!(m - 1)!]p^{h+1}q^{m-1}. \tag{5C.3}$$

Now, by formula (5.2) in the main text, in which h and m are replaced with $(h + 1)$ and $(m - 1)$, respectively, we find that the coefficient for the combination of $(h + 1)$ hits and $(m - 1)$ misses is given by:

$(h + 1 - 1 + m - 1)!/h!(m - 1)! = (h + m - 1)!/(h)!(m - 1)!$

which is identical to the coefficient within the brackets on the right-hand side of (5C.3). In other words, the sum $\Sigma[(h + m - 1)!/(h)!(m - 1)!] p^{h+1}q^{m-1}$ is equal to the sum of the probabilities of all possible combinations of $(h + 1)$ hits and $(m - 1)$ misses and therefore is equal to 1. Substituting this result in (5C.3), and noting that $q = 1 - p$, we find:

$$\mu = hq/p = h(1 - p)/p.$$

(ii) Calculation of the Variance $\sigma^2 = h(1 - p)/p^2$

The variance of the m is defined as the average (or more formally the expectation) of the squared deviation of m about the mean μ, i.e. Expectation$(m - \mu)^2$ or compactly $E(m - \mu)^2$. Then, inasmuch as μ was given by $\Sigma m \Pr(m)$ in (5B.2), σ^2 is given by $\Sigma(m - \mu)^2 \Pr(m)$; i.e.

$$\sigma^2 = \Sigma(m - \mu)^2 \Pr(m) = \Sigma m^2 \Pr(m) - 2\mu \Sigma m \Pr(m) + \mu^2 \Sigma \Pr(m). \tag{5C.4}$$

But, then, the last term $\mu^2 \Sigma \Pr(m) = \mu^2$ because $\Sigma \Pr(m) = 1$. Also, $2\mu \Sigma m \Pr(m) = 2\mu^2$ because, as shown in (5C.2), $\Sigma m \Pr(m) = \mu$. Substituting those in the right-end of (5C.4), we find:

$$\sigma^2 = \Sigma m^2 \Pr(m) - \mu^2. \tag{5C.5}$$

Now that $\Pr(m) = \Sigma[(h - 1 + m)!/(h - 1)!m!]p^h q^m$ as given in (5C.2), we find:

$$\Sigma m^2 \Pr(m) = m^2 \Sigma[(h - 1 + m)!/(h - 1)!m!]p^h q^m$$
$$= m\Sigma[(h - 1 + m)!/(h - 1)!(m - 1)!]p^h q^m, \tag{5C.6}$$

because $m/m! = 1/(m - 1)!$. I now manipulate $m = 1 + m - 1$. Furthermore, I manipulate the second right-hand side so as to decompose it into two parts:

$$(1 + m - 1)\Sigma[(h - 1 + m)!/(h - 1)!(m - 1)!]p^h q^m$$
$$= \Sigma[(h - 1 + m)!/(h - 1)!(m - 1)!]p^h q^m$$
$$+ (m - 1)\Sigma[(h - 1 + m)!/(h - 1)!(m - 1)!]p^h q^m$$
$$= \Sigma[(h - 1 + m)!/(h - 1)!(m - 1)!]p^h q^m$$
$$+ \Sigma[(h - 1 + m)!/(h - 1)!(m - 2)!]p^h q^m \tag{5C.7}$$

because $(m-1)/(m-1)! = 1/(m-2)!$ in the second term (with the second summation sign) of the first right-hand side. The first term on the right-hand side can be further manipulated in such a way that:

$$\Sigma[(h-1+m)!/(h-1)!(m-1)!]p^h q^m$$
$$= (hq/p)\Sigma[(h+m-1)!/h!(m-1)!]p^{h+1}q^{m-1} = hq/p.$$

This is because the summation on the second term of the right-hand side of (5C.7) is the sum of all probabilities of $(h+1)$ hits and $(m-1)$ misses, which is equal to 1.

The second sum of the right-hand side of (5C.7) is similarly manipulated to obtain:

$$\Sigma[(h-1+m)!/(h-1)!(m-2)!]p^h q^m$$
$$= [(h+1)hq^2/p^2]\Sigma[(h+2-1+m-2)!/(h+1)!(m-2)!]p^{h+2}q^{m-2}$$
$$= (h+1)hq^2/p^2$$

(5C.8)

because the summation in the first right-hand side is the sum of all probabilities for $(h+2)$ hits and $(m-2)$ misses which is equal to 1. Substituting the above results in (5C.5), recalling the previous result that $\mu = hq/p$, and noting that $p+q=1$, we finally find:

$$\sigma^2 = \Sigma m^2 \Pr(m) - \mu^2 = hq/p + (h+1)hq^2/p^2 - (hq/p)^2$$
$$= h(q/p)(1+q/p) = h(1-p)/p^2.$$

Appendix 5D: Why Do the Terms jk^{j-1} Qr(j) in (5.6) Vanish in the Limit ($j \to \infty$)?

First of all, notice that a Qr(j) is a proportion such that it takes a value only in the interval [0, 1], regardless of j. That is, it is no more than 1 for $j \to \infty$. So, I only need to show that the expression jk^{j-1} vanishes for $j \to \infty$. However, for $j \to \infty$, the expression k^{j-1} obviously vanishes because $0 < k < 1$ as defined. Thus, the entire expression jk^{j-1} becomes $\infty \times 0$ which, as it stands, is indeterminate. However, we know the l'Hôpital rule already shown in Appendix 4C. To repeat: Consider that both $f(u)$ and $g(u)$ tend to infinity in the limit $u \to a$ and that $f(u)$ and $g(u)$ are both differentiable at $u = a$, i.e. they have derivatives $f'(a)$ and $g'(a)$. Then:

$$f(u)/g(u) = f'(u)/g'(u) \text{ in the limit } (u \to a).$$

Let's see how his trick works in the present case. For this purpose, reset the original expression jk^{j-1} as $j/(1/k^{j-1})$, such that $(1/k^{j-1}) \to \infty$ for

$j \to \infty$. Thus, writing $j \equiv f(j)$ and $(1/k^{j-1}) = k^{1-j} \equiv g(j)$, we find both f and g tend to ∞ for $j \to \infty$. Then, we can evaluate $jk^{j-1} = f(j)/g(j) \to \infty/\infty$ as $j \to \infty$, using the l'Hôpital rule. The problem, however, is that u is continuous and j, integers. Nonetheless, we can justify considering j to be practically continuous because we are dealing with sufficiently large values of j. Then, in Appendix 4B, you find the derivatives $f'(j) = 0$ by the standard function (i), and $g'(j) = [-k^{(1-j)}] \times \ln(k)$ by (c) and (ii). Then, for $j \to \infty$, $g'(j) \to \infty$ because $0 < k < 1$. Thus, for $j \to \infty$, the expression $jk^{j-1} = f(j)/g'(j) = 0/\infty = 0$. That is, $jk^{j-1}Qr(j)$ vanishes in the limit $j \to \infty$.

Appendix 5E: Convergence of the Sum $\{\Sigma[(h + j - 1)!/h! (j - 1)!](kq)^{j-1}\}$ to $(1 - kq)^{-(h+1)}$

Recall model (5.3) for the trap-shooting game in the main text. To repeat:

$$p^h \Sigma[(h - 1 + m)!/(h - 1)!m!]q^m = 1. \quad (5.3 \text{ rpt})$$

Let us make some notational changes on the left-hand side without changing its structure (as the sum of the probabilities of all possible events) such that it remains equal to 1:

Original	Changed
h	$(h + 1)$
$(h - 1)$	h
m	$(j - 1)$
q	kq
p	$(1 - kq)$

Then, (5.3 rpt) is modified to:

$$(1 - kq)^{h+1} \{\Sigma[(h + j - 1)!/h!(j - 1)!](kq)^{j-1}\} = 1, \text{over} j = 0, 1, 2, \ldots, \infty.$$

Thus, multiplying both sides by $(1 - kq)^{-(h+1)}$, we find:

$$\Sigma[(h + j - 1)!/h!(j - 1)!](kq)^{j-1} = (1 - kq)^{-(h+1)}.$$

Voila! [Note: Changing the notation p to $(1 - kq)$ as above applies only to the present framework, i.e. the manipulation of (5.3 rpt) for the purpose of evaluating the particular summation concerned. The notation p in (5.10) remains as $(1 - q)$, i.e. it is not changed to $(1 - kq)$.]

6 · Scramble and Contest Competition: What Is the Difference?

6.1 Preamble

In his vast work 'An outline of the dynamics of animal populations', the Irish–Australian entomologist A. J. Nicholson (1954) distinguished two types of competition among individuals of a single-species population: 'scramble' and 'contest'. Essentially the distinction is as follows. Consider that the individuals of a population are more or less equally competitive and, as a result, the resources are divided more or less evenly among them. Then, as the population increases, a point would sooner or later be reached at which few individuals could get enough resources to survive and reproduce. So, eventually, all would go down together. This is 'scramble' competition. In contrast, some (usually stronger) individuals may outcompete the (weaker) others and manage to survive and reproduce. This is 'contest' competition. Altogether, the distinction between the two types of competition sounds so clear and simple, does it not?

Nicholson's idea quickly spread among ecologists. As straightforward as it may sound, the idea was not well perceived by many, often resulting in inappropriate interpretations, including, admittedly, my own. In the present chapter, I attempt to clarify the situation, and then to expand the idea into a principle for pragmatic use. To do it, I look at the idea quantitatively in the light of model (5.12) of competition developed in Chapter 5. In particular, I look into its reproduction curves, a method developed in Chapter 4. This allows me to perceive the distinction between the two types of competition primarily in terms of the variation in the parameter h of model (5.12). This parameter was originally defined as a measure of the degree of aggregation of individuals over their effective habitat space. So, it is closely related to how the resources are divided among them and how they compete. However, this is a rather restricted concept of h, and I shall in due course expand it to a measure of unevenness in competitiveness among the individuals of a population.

To make the following arguments simple, I assume that the environment is more or less homogeneous, or the resources are more or less evenly distributed over the effective habitat of the species concerned. In other words, I assume that the uneven share of resources among the competitors depends mainly on uneven competitiveness among them, rather than on the heterogeneity of the environment. To begin, let us draw a few reproduction curves, using model (5.12).

6.2 Drawing Reproduction Curves Based on Model (5.12)

First, I briefly review model (5.12). I reproduce it here, as well as its limiting case, i.e. model (3.10) for $h \to \infty$:

$$r_t = r_m/[1 + (1 - k)sx_t/h]^{h+1} \tag{5.12 rpt}$$

$$r_t = r_m \exp[-(1 - k)sx_t] \tag{3.10 rpt}$$

in which $r_t = x_{t+1}/x_t$ is the net rate of change in population density x from generation t to $t + 1$; r_m is the potential (maximum) reproductive rate of an individual when it has no competitors; s is the size of the area containing the minimum sufficient amount of resources for survival and reproduction; $0 < k < 1$ is a measure of the intensity of competition (given s) with another individual (the lower the k-value, the more intense); and $h > 0$ is, for now, a measure of the level of aggregation by the members of the population (the lower the h-value, the more aggregated): conversely, the limiting case (3.10) for $h \to \infty$ implies that the individuals are distributed completely at random, i.e. little aggregated.

Now, to draw reproduction curves in the form of Figure 4.3a from Chapter 4, I transform (5.12) and (3.10) into (natural) logarithms, using the notations $R = \ln(r)$ and $X = \ln(x)$. Also, I may write $(1 - k)s$ compactly as c. So, we have:

$$R_t = R_m - (h + 1)\{\ln[1 + (c/h)\exp(X_t)]\} \tag{6.1a}$$

$$R_t = R_m - c \exp(X_t). \tag{6.1b}$$

In the following, I arbitrarily fix the potential reproductive rate R_m at a certain value. This is because R_m is the parameter that would be realized when the individual concerned has no competitors and, hence, is not directly relevant to the present investigation into competition.

122 · **Scramble and Contest Competition**

6.2.1 Reproduction Carves for $h \geq 1$ and Finite $c > 0$

I suggest that we draw two reproduction curves, marked A and B in Figure 6.1. Curve A is based on model (6.1a) with the parameters set: $R_m = 2$ and $c = h = 1$. Curve B is based on (6.1b) in which $R_m = 2$, $c = 5$ and $h \to \infty$. These values are chosen for easy visual comparison. Curve B is arbitrarily positioned, and curve A is positioned very close to (but never below) curve B. Thus, the two curves are practically identical to start at a sufficiently low value of X_t but, as X_t increases, curve A decreases more slowly (less steeply) than B.

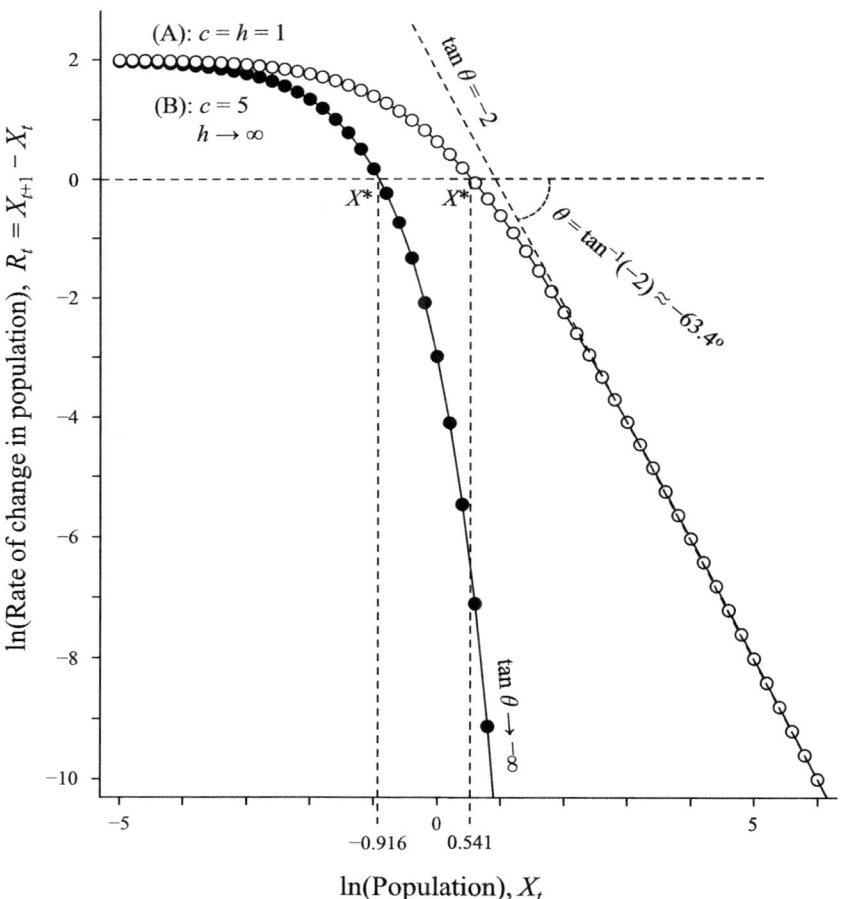

Figure 6.1 Graphical analysis of reproduction curves in visualizing the difference between two types of competition.

6.2 Drawing Reproduction Curves · 123

Now, Figure 6.1 gives visual impressions that, as X_t increases, curve A seems to converge asymptotically to a fixed tangent line, whereas curve B appears to plummet. The difference is ecologically significant in that it provides a theoretical criterion for distinguishing the two types of competition. To see it, though, visual impressions are not good enough. So, the calculation of the derivative of each equation in (6.1), i.e. dR_t/dX_t as a measure of the slope of a curve, is in order.

6.2.2 Calculation of the Derivatives of Models (6.1)

As already explained in Appendix 4A in Chapter 4, the slope of a curve (at a given point) is defined as the slope of the tangent line, i.e. ($\tan \theta$), which is measured by the derivative of the curve at that point, i.e. $\tan \theta = dR_t/dX_t$, given X_t. So, using the procedures of calculation in Appendix 4B, we find the derivatives of (6.1a, 6.1b) to be, respectively:

$$dR_t/dX_t = -(h+1)\{[(c/h)\exp(X_t)]/[1+(c/h)\exp(X_t)]\} \quad (6.2a)$$

$$dR_t/dX_t = -c\exp(X_t). \quad (6.2b)$$

The visual impression that curve B (in Figure 6.1) plummets (as X_t increases) is readily confirmed by the right-hand side of (6.2b) tending to negative infinity, i.e. the slope of the curve for a large X_t tends to become vertical downwards. [Note: Recall in Chapter 4 that a negative derivative means a downwards slope.]

In contrast, the slope of every curve generated by (6.1a) with a finite h, i.e. derivative (6.2a), tends to $-(h+1)$ as X_t increases: that is, in the limit $X_t \to \infty$, $\tan \theta \to -(h+1)$, or $\theta = \tan^{-1}[-(h+1)]$ in radian. This is because, as X_t increases, given $(c/h) > 0$, the quotient in the braces on the right-hand side of (6.2a) sooner or later converges to 1: because it is of the (generic) form $u/(1+u)$, it converges to 1 as u increases indefinitely. In curve A, $h = 1$ as assumed, and hence $\tan \theta = -(h+1) = -2$. Then, a click on your pocket calculator shows: $\tan^{-1}(-2) \approx -63.4°$. So, we see that curve A asymptotically converges to the slanted line that intersects a horizontal axis at angle $-63.4°$. In general, the slope of a reproduction curve generated by (6.1a) would become steeper as h increases but would never plummet as long as h stays finite.

[Note: As Appendix 4D of Chapter 4 has shown, the parameters R_m and c control the position of a reproduction curve generated by (4.4b), replicated as (6.1b) here. In particular, R_m controls the vertical position, whereas c controls the lateral position, of the reproduction curve in the (R_t, X_t)

124 · Scramble and Contest Competition

coordinate space. In light of (6.2a), we see that the additional parameter h controls the curvature of a reproduction curve generated by (6.1a).]

6.2.3 Reproduction Curves to Depict the Two Types of Competition

What does the difference in shape between the two curves in Figure 6.1 signify in terms of the two types of competition? To see it, I suggest that we examine what each curve does in the vicinity of its equilibrium point X^*. So, let us algebraically evaluate X^* first.

By definition, at the equilibrium X^*, it must be that $R_t = 0$ (i.e. no change in the X_t from the t-th generation to the next). So, substituting $R_t = 0$ and $X_t = X^*$ in (6.1a), and after a little rearrangement of terms about the equal sign, we find:

$$R_m = (h+1)\{\ln[1 + (c/h)\exp(X^*)]\}$$

and solving the above for X^* (your homework), we have:

$$X^* = \ln\{\exp[R_m/(h+1)] - 1\} - \ln(c/h). \quad (6.3a)$$

Likewise, we find in (6.1b):

$$X^* = \ln(R_m/c). \quad (6.3b)$$

Because the parameter values are ($R_m = 2$, $c = h = 1$) in curve A, substituting these values in (6.3a) we find $X^* = \ln\{e^{2/2} - 1\} - \ln(1/1) = \ln(e - 1) - 0 \approx 0.541$, as marked on the horizontal X_t-axis in Figure 6.1. In curve B, the parameter values are ($R_m = 2$, $c = 5$), and hence $X^* = \ln(2/5) \approx -0.916$.

Now, exactly at X^*, the same number of offspring is produced (at the same developmental stage) as the parental population in each curve, i.e. $X_{t+1} = X_t = X^*$. But what if the parental population (X_t) was displaced from X^* to the right by a given distance? Suppose that the distance of displacement is 2 units above X^* in each curve. We have the following results:

	Curve A	Curve B
X^* by (6.3)	$\ln(e-1) \approx 0.541$	$\ln(2/5) \approx -0.916$
$X_t = X^* + 2$	2.541	1.084
R_t in (6.1)	−3.234	−12.778
$x_{t+1}/x_t = \exp(R_t)$	0.039	0.0000028

The bottom row for x_{t+1}/x_t shows that, after the same unit of displacement (i.e. $X_t = X^* + 2$), the number of offspring (x_{t+1}) produced in curve A is about 4% of the parental population (x_t), whereas in curve B it is as low as 0.0003%. What if we increased the distance of displacement even further? The (ln) reproductive rate R_t decreases in both curves. However, in curve A, the rate decreases only at an arithmetic rate (because the curve is asymptotically linear, i.e. a straight line), whereas it decreases exponentially in curve B. The following is the ecological significance of these results in terms of competition.

6.2.4 Curves A and B as Respective Examples of Contest and Scramble Competition

Recall that curve B represents the situation in which the spatial distribution of individuals is least aggregated over their effective habitat space, indicated by $h \to \infty$. Then, at a certain level of population size, and as long as the environment is reasonably homogeneous, the available resources would be more or less evenly divided among all competing individuals. Now, as X_t increases, the average reproductive rate decreases at an exponential rate in curve B, and we see that, sooner or later at a higher population level, practically no offspring would be produced. Thus, *we may* say that the curve B situation quantitatively exemplifies the notion of scramble competition as verbally stated at the beginning of this chapter. [Note: The reason for the italicized phrase *we may* will become apparent shortly.]

In contrast, curve A (with $h = 1$) represents a situation in which individuals are unevenly distributed spatially, and some individuals manage to secure more resources than other individuals. If this uneven distribution is the result of competition, i.e. the stronger individuals outcompete the weaker ones to secure enough resources, *we may* consider that curve A fits the notion of contest competition. [Note: An uneven distribution of individuals could result from a heterogeneous environment in which some lucky ones found themselves in better patches or refuges without competing. But we are talking about competition over a more or less homogeneous environment. So, the effect of environmental heterogeneity is outside the scope of the present theme, although I shall come back to this issue later.]

The above perception notwithstanding, we cannot draw a sharp line between curves A and B because, mathematically, the values of h change continuously from a given value to another: even a change from a finite value to infinity is continuous. In other words, no sharp line can be drawn between the two types of competition inasmuch as the transition from curve A to B is a continuum. That is, we cannot pick up a given curve and say: this represents scramble competition and that, contest. This is why I used the expression 'we may' in the foregoing paragraphs.

The bottom line is this: although a change from one curve to another is continuous, if you pick two curves adequately separated from each other, as in Figure 6.1, we would see the difference. With this perception in mind, I continue investigating the nature of competition. To begin with, I broaden the meaning of the parameter h.

6.3 Broader Interpretation of Parameter h

First of all, rather than restricting the parameter to a measure of spatial aggregation of the individuals, we may generalize it to a measure of unevenness in competitiveness among individuals: the lower the h-value, the higher the degree of unevenness, and vice versa. The unevenness could be due to differences in age (e.g. the older the stronger and vice versa, even though I am getting weaker every moment at my age) or due to a difference in genetic traits. In the following, I look into the effect of changes in h value on the shape (curvature) of the reproduction curves that model (6.1a) generates. However, we have already seen the effect of h changing from infinity (curve B) to 1 (curve A) in Figure 6.1. So, let us look into reproduction curves in the range $1 > h > 0$ in the following.

6.3.1 Reproduction Curves for $1 > h > 0$

If h is close to 1, the shape of a curve in this category would not be much different from curve A in Figure 6.1. So, why don't we go straight to the other extreme, $h \to 0$? I can think of two different situations in terms of model (6.3a).

Situation (i): c is a positive constant independent of h. Then, the quotient (c/h) on the right-hand side of (6.3a) becomes infinitely large, resulting in $X^* \to -\infty$, i.e. population extinction. Evidently, this is outside our interest: simply, there would be no competition among nobody. Situation (ii): $c \to 0$ as $h \to 0$. Then, mathematically, the quotient c/h can assume some positive finite value (for instance, if $c = h = 0.00000...01$, then $c/h = 1$), and hence X^* remains finite. In fact, for $R_m = 2$ and $c = h \to 0$, we find $X^* \approx 1.855$ by equation (6.3a); and by equation (6.2a), the slope of the curve converges to $\tan^{-1}(-1) = -45°$

as X_t increases, compared with $\tan^{-1}(-2) \approx -63.4°$ in curve A of Figure 6.1. But what does ($c \to 0$, $h \to 0$), written compactly as ($c, h \to 0$), mean ecologically?

6.3.2 Ecological Interpretation of ($c, h \to 0$)

Recall the equivalence $c \equiv s(1 - k)$. Then, algebraically, $c \to 0$ means either $s \to 0$ or $k \to 1$. But, s is the size of a quadrat, and it is ecologically meaningless to consider $s \to 0$. Thus, we must consider the alternative, i.e. $k \to 1$. However, certain conceptual problems arise.

The parameter k (as a measure of the intensity of competition) is defined in the interval $0 < k < 1$: the larger the k-value the less intense is competition. Thus, literally interpreted, $k \to 1$ implies little or no competition. Then, to make the quotient c/h (in which $h \to 0$) to be positive and finite-valued, we may have to assume that little (or no) competition is involved in the process concerned, but this creates a logical problem.

Notice that, for $c = h \to 0$, the slope of the reproduction curve converges to $-45°$ as was mentioned in the preceding section. It indicates that this reproduction curve is not much different in shape from curve A in Figure 6.1, in which $c = h = 1$. But, as we have interpreted, curve A is supposed to exemplify contest competition. How is it possible that the process, in which no competition is involved, resembles curve A? Or more precisely, how come the rate of change in a process with no competition still decreases as the population increases? In fact, the problem lies in the original perception of the ecological meanings of h and k as explained below.

So far, h has been considered to be a positive parameter. But, if we insist on this original definition, $h \to 0$ should be an extreme case of the reproduction curves that model (6.1a) can generate, given (c/h) > 0 and finite. In other words, if restricted to $h > 0$, the model would not generate a reproduction curve whose steepest slope is less than $45°$ in absolute value. However, there is no reason to consider that natural populations in general are so restricted. As a matter of fact, the Hassell model – which has been shown (in Chapter 5) to be a descriptive version of my model (5.12) and is equivalent to (6.1a) after ln-transformation – actually produces more varieties of reproduction curves beyond the range $0 < h$, and these curves do generate the corresponding varieties of population series that look very realistic. In other words, as a descriptive device, the Hassell model is more flexible (and realistic) than model (5.12). The reason for the difference in flexibility is as follows.

128 · **Scramble and Contest Competition**

The Hassell model contains the parameter b defined as a positive constant, i.e. $b > 0$. However, a comparison between the Hassel model and model (5.12) in Chapter 5 has shown $b = (h + 1)$, such that $h > 0$ implies $b > 1$. This means that, for $h > 0$, my model (6.1a) excludes the potential range $0 < b \leq 1$ of the Hassell model. That is, (6.1a) excludes the range $0 < (1 + h) \leq 1$, i.e. $-1 < h \leq 0$.

Altogether, for model (6.1a) to comply with the realism of the Hassell model, the range of variation in h has to be extended to the negative domain, i.e. analogous to a negative hit in trapshooting. Weird as it may sound, it does provide a deeper insight into the nature of competition. After all, we have generalized h to be less than 1 (i.e. a fraction of a hit). So, we might just as well generalize it to be negative, and investigate the nature of this weird thing. Remember that, as I have pointed out previously, some weird mathematical abstractions may provide insight. So, let's go!

6.4 In the Weirdland of a Negative Hit

OK, I suggest that we follow Alice to Weirdland to meet Miss Negative Hit who can turn h into a negative number. What happens? Take a look at the equilibrium point X^* in (6.3a), as repeated below:

$$X^* = \ln\{\exp[R_m/(h+1)] - 1\} - \ln(c/h). \quad \text{(6.3a rpt)}$$

For $h < 0$ and $c > 0$, the quotient $(c/h) < 0$ and the term $\ln(c/h)$ on the right-hand side becomes a complex number. That is, in the real world, we cannot attach any ecological significance to a negative h when $c > 0$. [Note: The logarithm of a negative number is a complex number. If you are curious about this, see Appendix 6A.] So we ask Miss N. Hit: 'What shall we do, Miss?' With a big grin, she shows us a solution, in fact an insightful one. Let us first look into the range $-1 < h < 0$.

6.4.1 Case for $-1 < h < 0$: Sociality vs. Competition

Her trick is to broaden the original definition of $0 < k < 1$, a measure of the intensity of competition: to repeat it again, the closer k is to 1, the less intense is competition, and vice versa. Here, we expand k to be larger than 1. Then, $c = s(1 - k)$ becomes a negative number so that the quotient (c/h) stays positive, and the X^* in (6.3a) remains real mathematically.

But, ecologically, how can we interpret $k > 1$ (and hence $c < 0$) on the condition that $-1 < h < 0$? The answer is: sociality (an individual benefit of aggregation or even cooperation) as opposed to competition. That is, forming a school, shoal, flock, gaggle, herd, or a socially more organized (orderly) troop, pack, etc. (excluding a mob) may increase the fitness of an individual. Thus, $k > 1$ (and hence $c < 0$) implies the formation of social groups, causing a highly aggregated distribution of individuals in the population, indicated by $-1 < h < 0$.

Keep in mind, though, that it is not reasonable to assert that $k < 1$ means competition and $k > 1$, sociality: such a black-and-white perception may mislead. The process of a change from one state to the other should be continuous. But, then, there is an apparent problem: in transition from $k < 1$ to $k > 1$ (i.e. from positive to negative in c) there must be the point at which $c = 0$, i.e. $k = 1$. Then, for $h < 0$, the quotient $(c/h) = 0$ and the term $-\ln(c/h)$ on the right-hand side of (6.3a) becomes infinity and so does X^*. That is, the population would not exist in the real world. However, c can be arbitrarily close to zero in transition from positive to negative values, and vice versa. Then, the quotient (c/h) can, in its limiting form, be finite and positive no matter how small c and h may become in absolute values, so long as they are of the same order of infinitesimal with the same sign. This ensures the existence of the population. [Note: The foregoing consideration is necessary to maintain mathematical rigour because ecological feasibility depends on mathematical feasibility, if not vice versa.]

Thus, after all, $k \to 1$ (or $c \to 0$) can evolve if associated with an appropriate value of h such that the quotient (c/h) is positive and finite-valued. On this basis, I interpret $(c, h \to 0)$ in particular to imply that the benefit of aggregation and the counter effect of competition are in a balanced state, so to speak. It follows that, on the one side of this balanced state (where $k < 1$, hence $c > 0$, and $h > 0$), the effect of competition is greater than the benefit of sociality, and vice versa on the other side (where $k > 1$, hence $c < 0$, and $h < 0$): the benefit of sociality outweighs the effect of competition.

But, what if k was so large that c/h became large? The equilibrium point X^* in (6.3a) would accordingly become small, a situation which I interpret as follows. If the majority of individuals in the population become too sociable in pursuit of each own benefit gained from being sociable, they would begin to hurt each other: or more importantly, each individual involved would be in trouble. Then, as a result, the per-capita reproductive rate of each individual would begin to decrease. In other words, too large a k value (as a genetic trait of being excessively sociable) implies a self-inflicting disaster and is unlikely to evolve. Thus, there must be an equilibrium state between the advantage and disadvantage of gathering, or k must be bounded somewhere from above; i.e. there should be an upper limit.

Altogether, by introducing the notion of sociality, indicated by $k > 1$ and $-1 < h < 0$, we can extend the repertoires of reproduction curves generated by (6.1a) to represent all these situations, ranging from more or less purely individual-level competition to an organized social gathering (group, community, etc.). Now that we have come this far, we might just as well take one further step into the Weirdland of negative hits. That is, to consider the extreme case of $h \to -1$, or $h = -1$ practically, and hopefully no further.

6.4.2 $h = -1$ Means Malthusian

First of all, for $h = -1$, c must stay negative to avoid an ecological infeasibility of the logarithm of a negative number as already discussed. It is also incumbent that $R_m > 0$, or $r_m > 1$, for the population concerned to avoid its inevitable extinction. Under these constraints, the substitution of $h = -1$ or $(h + 1) = 0$ in model (6.1a) yields $R_t = R_m$, or $r_t = r_m$, i.e. the rate of change in the population is positive and constant. What happens? The population increases exponentially with time, implying the Malthusian increase in population or a self-inflicting disaster.

Here I stop carrying on any further with negative h because, beyond this line, I see nothing relevant to ecology in the real world. Instead, let me summarize what we have done thus far about the nature of competition.

6.5 Nature of Competition

The effect of differences in competition types is exemplified in the variation in shape (curvature) of a reproduction curve generated by models (6.1). The ecological mechanisms underlying these reproduction curves are generalized to include sociality to counter the effect of competition. The resultant variation in reproduction curves is graphically summarized in Figure 6.2. All curves share the same value of R_m, arbitrarily fixed at 2 as in Figure 6.1.

The curves are arranged in the order of the value of h, starting from the left with one for $h \to \infty$ as an extreme case of competition,

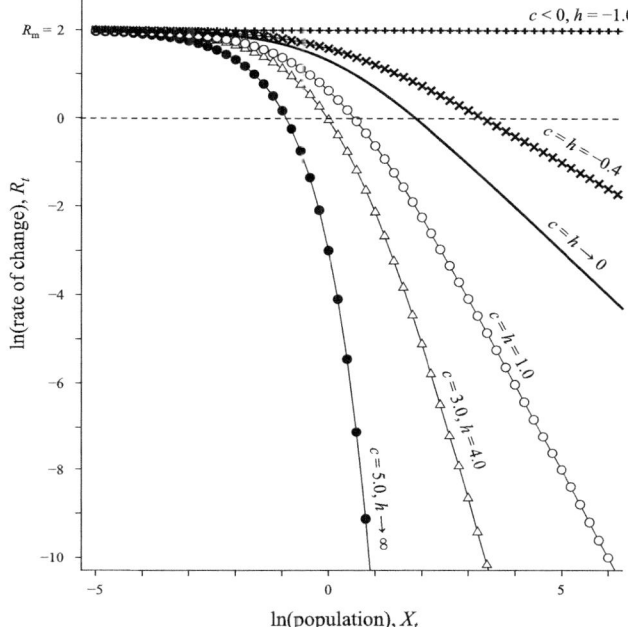

Figure 6.2 Variations in reproduction curves in accordance with the h values in model (6.1).

i.e. model (6.1b), all the way to the right and up, ending with one for $h = -1$, representing no competition (or Malthusian). In between the two extremes, there is the (plain bold) curve for $(c = h \to 0)$ as a particular case of $(c, h \to 0)$. This serves as a reference curve: on its left, the effect of competition overrides the effect of sociality, and the other way around on the right.

It is important to recognize that all these curves vary as a continuum, such that a sharp line cannot be drawn between competitiveness and socialness. Therefore, we should not consider the reference curve $(c = h \to 0)$ as the border separating competition and sociality. We may only say that the reference curve represents a situation in which the effects of competition and sociality are in balance, so to speak. In other words, those curves below (left of) the reference curve (i.e. those with $h > 0$) are inclined to competition (or a higher content of competitive elements), as opposed to those above the reference line ($h < 0$) that are inclined more towards sociality (or a higher content of social elements). This recognition suggests that those curves above the reference curve do not carry information in a pure form about the element of competition, let alone information about its types: only those below the reference curve (i.e. those with $h > 0$) may do so. [Note: The theme of the present chapter is the nature of competition. The nature of sociality is an important subject in its own right, but I never studied it in my active carrier. So, this is beyond the present book.]

However, even among those curves below (left of) the reference curve, the two types of competition cannot be sharply distinguished because their variation is a continuum. We may only say that a high value of h (within its positive range) implies a higher content of the scramble element, as opposed to a lower h indicating a comparatively higher content of the contest element. More precisely, we may consider the curve for $h \to \infty$ to exemplify an extreme form of competition (the scramble type), whereas we may not find a curve that typifies contest competition. This is because, as the h-value becomes lower, we can only say that the proportion of the contest element increases. Nonetheless, as shown shortly, it is unlikely that the contest element would completely displace the scramble element. Furthermore, before the element of contest competition becomes predominant, the elements of sociality may come into effect to obscure the effect of competition.

With the above thought in mind, I invite you to take a look at what I have found in some laboratory experiments. The main issue here is to look into what determines the parameter h of the model process (6.1a).

6.6 What Determines Parameter h in Actual Processes?

As I referred to in Chapter 5, the late Professor Shunro Utida of Kyoto University performed two types of experiments with azuki bean weevil: discrete and series experiments. [Note: The results of his experiments have been described and analysed in my previous book (Royama, 1992, pp. 237–65).] Here, I show that the two experiments differ from each other in the estimated values of h, revealing their difference in the mode of competition. Let me first describe the discrete experiment.

6.6.1 Discrete Experiment

In this type of experiment, Utida introduced varying numbers of weevils as breeding pairs into separate containers (Petri dishes, each containing 20 grams of beans) under a controlled temperature–humidity

6.6 What Determines Parameter h?

combination and let them lay eggs. After several weeks, their offspring began to emerge as second-generation adults in each container, and he counted them.

The initial number of breeding pairs in each container was varied as {1, 2, 4, 8, 16, 24, 32, 48, 64, 96, 128, 192, 256, and 384} with up to 10 replicates: there were four separate sets of experiments, of which I use the results shown in graph (d′) of figure 7.2 in Royama (1992). Letting x_1 be twice the number of breeding pairs in each container, and x_2 be the (mean) number of their emergent (adult) offspring, I calculated the log (mean) reproductive rate of a breeding pair, i.e. $\log_{10}(x_2/x_1) = (X_2 - X_1) = R_1$. The calculated R_1 is regressed on X_1 in Figure 6.3a (series of solid circles) to form an observed reproduction curve. Furthermore, using the observed results, I estimated the parameters (R_m, h, c/h) of model (6.1a)

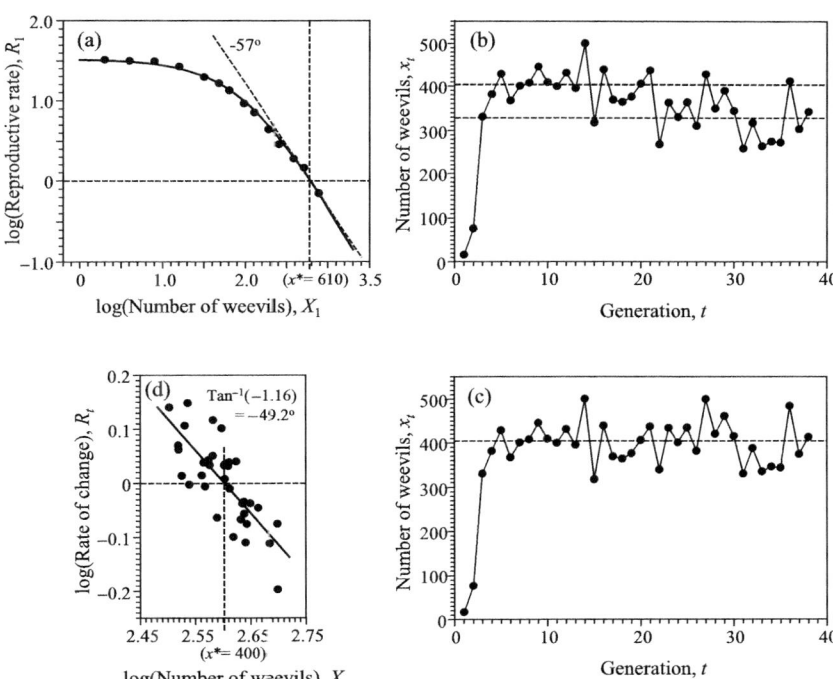

Figure 6.3 Results of the Utida experiments with azuki bean weevil. (a) The result of the discrete experiment fitted with model (6.1a). (b) The result of the series experiment, with adjustment in (c). (d) Estimation of h value in the series experiment.

and calculated a theoretical reproduction curve (the smoothed curve in Figure 6.3a) to fit the observed points. [Note: Because I am dealing with actual observations here, the results in Figure 6.3a are plotted in the $\log_{10}$ scale.] Appendix 6B describes the details of the estimation procedure. We see that the fit is very good, implying that the estimates $(R_m, h, c/h) = (1.52, 0.8, 0.01)$ explain the observed result with little error.

6.6.2 Series Experiment

In this type of experiment, Utida introduced eight founder pairs ($x_1 = 16$) into a single Petri dish (with 10 grams of beans) to establish a population, and let it continue for a number of generations. The population was maintained by periodically replenishing beans and removing debris. Utida counted the number of offspring weevils (emerged as adults) weakly and estimated the number of adults in each generation. Figure 6.3b shows the series of those estimated numbers of emergent adults plotted against generation. There were four series (see figure 7.11, p. 260, in Royama, 1992) with similar results, but I chose one (graph d of figure 7.11) that suits my purpose here in Figure 6.3.

To evaluate the h-value in this series experiment, let us take a look at equation (6.2a rpt) in Appendix 6B. We see that we must know R_m and $\tan \theta$ to evaluate h, but the series experiment provides no information to estimate these. However, we have already estimated $R_m = 1.52$ in the discrete experiment. This can be used as an estimate for the series experiment because, being the potential reproductive rate when there is no competitor, there is no reason to think that R_m should differ between the two experiments, performed under the same environmental conditions. Hence, we only need to estimate $\tan \theta$, the slope of the reproduction curve at X^* in the series experiment. This can be readily done in the following way.

Notice first that the observed series in Figure 6.3b has reached an apparent equilibrium level by $t = 3$. Then, using the section after $t = 3$, we should be able to calculate the log reproductive rate $\log(x_{t+1}/x_t) \equiv (X_{t+1} - X_t)$, and regress it on X_t. The slope of the regression line should estimate the target value of $\tan \theta$. However, I see a small technical problem here. The observed series appears to change (for an unknown cause) its equilibrium levels at about $t = 22$: it is at 400.5 from $t = 3$ to 21, but thereafter somehow shifts down to 329.1. Then, the raw data are no good for the regression. The simplest solution to get around the problem is to equalize the two sections of the observed series by shifting the second section up to the level of the first as much as 71.4 as in

Figure 6.3c. So, the series now has a single equilibrium level at $x^* = 400.3$. [Note: Too much manipulation, you might say? If I could replicate the experiment, I would have. The reality is that this sort of information is so scarce that I justify the manipulation.]

Let us now calculate the ($\log_{10}$) reproductive rates ($X_{t+1} - X_t$), using the series Figure 6.3c for $t = 3, 4, 5, \ldots, 38$. Each of the calculated rates is then regressed on the corresponding X_t. The result is shown in Figure 6.3d. We see that the regression line goes neatly through the observed equilibrium $\log_{10}(400.3) = 2.60$ at which $R_t = 0$. In other words, the slope of the regression line (the regression coefficient) is a good approximation of the $\tan \theta$ of the (unknown) reproduction curve of the series experiment at the equilibrium point X^*. The regression coefficient turns out to be -1.16. Thus, substituting $R_m = 1.52$ and $\tan \theta = -1.16$ in the equation (6.2a rpt) in Appendix 6B, and after a little algebraic iteration, I find $h = 0.2$ in the series experiment. This estimate is substantially different from the estimate $h = 0.8$ in the discrete experiment. So, let us look into the ecological significance of the difference.

6.6.3 Interpretation of the Difference in h-Values between Discrete and Series Experiments

When I was writing my previous book (Royama, 1992), I held the following view. In the discrete experiment, all the weevil females in a given rearing container laid eggs more or less at the same time. Consequently, the eggs hatched more or less at the same time. This must have promoted scramble competition among the larvae. In contrast, in the series experiment, the age of the larvae must have spread out after a few generations, and this must have promoted contest competition, i.e. older larvae have competitive advantage over younger ones.

I now realize that, in the light of the foregoing analysis, my earlier perception was not quite right: around that time, I did not have the competition model (5.12), let alone the parameter h. As is clear in Figure 6.2, the estimated $h = 0.8$ in the discrete experiment does not appear to be large enough to imply a high content of scramble element: 0.8 is even lower than $h = 1$ in curve A of Figure 6.1, which suggests that the tendency is inclined more towards contest competition. After a while, though, I found the following fact which explained the above problem.

Even in the discrete experiment, each female in fact laid her eggs over a week or 10 days before she passed away. So, among the offspring larvae, there could have been 10 age-classes (in days) with different levels of competitive advantage. Considering that the average length of the larval stadium was about 25 days, the difference in age as much as 10 days must have promoted contest competition between older and younger age classes. This explains why the estimated h value ($0.8 < 1$) is rather small, compared with a much larger value as would have been expected were the eggs laid over a much shorter period of time, say within just a day or two. In other words, it was likely that, within each age-class, the larvae tended to compete in a scramble manner, but between age-classes in a contest manner.

Thus, the difference in the h-value between the discrete and series experiments is attributable to a difference in the spread of age-classes: longer in the series than in the discrete experiments.

Incidentally, the estimate $h = 0.2$ in the series experiment indicates an inclination towards contest competition rather than an involvement of sociality as it is unlikely to happen in a weevil population. [Note: there is one more problem in the observed reproduction curve in Figure 6.3a: there is only one data point past X^*. In other words, the crucial section of the curve about X^* did not have a good degree of resolution for estimating the tan θ at X^*. I consider that this is a minor problem in that the curve fits so well for the rest of the data points.]

The foregoing analysis of the laboratory experiments confirms the perception that scramble and contest are the two elements of competition that could occur in varying proportions. In the following I consider how this perception applies to more general situations in nature.

6.7 Scramble and Contest as Elements of Competition

Before getting to the main issue, however, I will review briefly certain views held by some earlier authors, including myself.

6.7.1 Earlier Views

As I have discussed in Chapter 5, Hassell (1975) used the model

$$r_t = x_{t+1}/x_t = r_\mathrm{m}/(1 + ax_t)^b$$

to show that: 'In terms of the parameter, b, the condition $1 > b > 0 \ldots$ represents different degrees of contest alone, while the condition $\infty > b > 1 \ldots$ represents varying combinations of scramble and contest'.

I held a similar view, and even expanded it, in my previous book (Royama, 1992, pp. 150–2), and so did Brännström and Sumpter (2005) in their figure 3. I now realize that, in the light of what I have conceived in the foregoing analyses, these earlier views should be reconsidered and amended.

6.7.2 Reconsideration and Amendment

To aid my contention (against my older view), I created Figure 6.4 to replicate the R_t-vs.-X_t reproduction curves of Figure 6.2, but in the x_{t+1}-vs.-x_t form (as in the original Moran plot in Chapter 4): I created Figure 6.4 because you likely encounter this form and its variants in the ecological literature. The point I attempt to amend here is the following.

In terms of model (5.12), Hassell's (1975) parameter b is equivalent to $(h + 1)$ and, hence, the condition $(1 > b > 0)$ is equivalent to $(-1 < h < 0)$. This is the situation represented by the reproduction curves in Figure 6.4a. However, in these curves, as I have shown in the Figure 6.2 version of the reproduction curves, the effect of competition would in general be eclipsed by the involvement of sociality (an individual benefit of aggregation). In other words, in the Figure 6.4a situation, we would not see the effect of competition clearly, let alone its elements (scramble and contest). Now, the situation $\infty > b > 1$ (in Hassell, 1975) is equivalent to $0 < h < \infty$ in (5.12), which is represented in Figure 6.4b where we begin to see the effect of competition for sure. Then, I agree with Hassell's perception that both elements are involved in varying degrees.

But how could we quantify the 'varying degrees'? It would be hard to do, to tell you the truth. Nonetheless, the following is the way I actually see the nature of competition that is made up of the two elements and that can occur even within a social group.

6.7.3 Nature of Competition as I See It

Let me begin with an example. In a temperate region, many species of birds start breeding more or less simultaneously at a specific time of the year. Often, this causes intense competition among them in establishing (breeding) territories prior to building nests. However, eventually, stronger (usually older) individuals stake out their territories in favourable

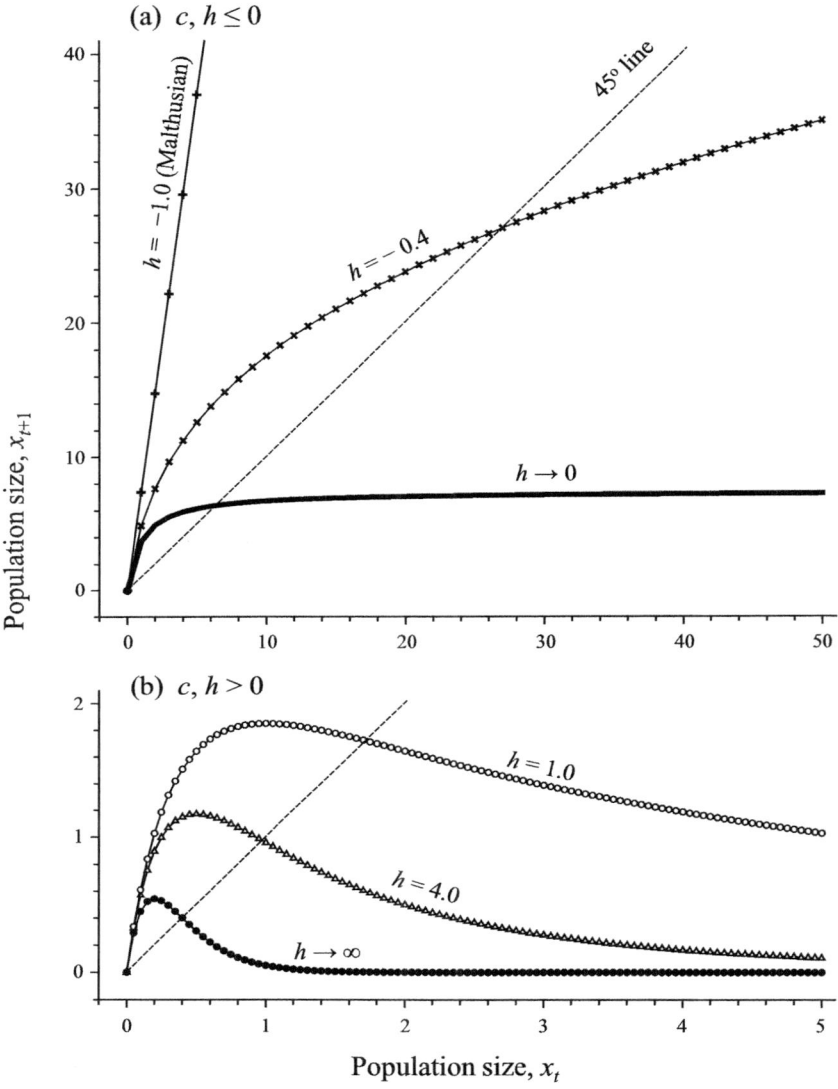

Figure 6.4 Reproduction curves in Figure 6.2 transformed back to linear-scale representations.

areas: a weak individual may not even be able to secure one and may remain in a non-breeding status. My interpretation of this process is as follows. Competition starts in a 'scramble manner' in that it starts on an equal basis among the competitors. However, as unevenness (inequality)

6.7 Scramble and Contest as Elements of Competition · 139

in competitiveness becomes apparent sooner or later, they compete in a 'contest manner'.

Another example is asynchronous hatching of eggs in a clutch. The siblings that hatch at more or less the same time might become involved in competition in the scramble manner, but between the earliest (stronger) and the latest (weaker) hatchlings, there could be competition in the manner of contest. It is well known among ornithologists that the youngest chick often would not survive. It is important to recognize in this example that we may see the effect of competition, no matter whether it is in the manner of scramble or contest, only when food was in short supply: if plentiful, all siblings would get an adequate share to survive.

Also, there are seasonal changes in the way these elements are involved: from the involvement of a competitive element for securing territories in the breeding season to the formation of a foraging flock outside the season. But, even within the foraging flock, as a manifestation of sociality, competition is involved. Right at this moment, I see a flock of juncos (*Junco hyemalis*) foraging on the ground just outside my study. Almost constantly, I see some individuals supplanting others to occupy a (seemingly) better feeding patch, although I see no particular sign of a hierarchical (pecking) order.

The recognition that competition is made up of two elements can be extended to those processes involving sociality, in that even in the Figure 6.4a situation where sociality is involved (as a dominant element) there must be elements of competition. Evidently, even within a social group (flock, herd, pack, troop, etc.), the element of contest competition is involved to establish social hierarchy. In the meantime, an element of aggressive competition is involved in intergroup relationships in varying degrees: violent fighting may occur among packs of wolves to secure their own territories in a scramble manner, whereas some primate troops may stay aloof from each other to avoid serious conflict.

Altogether, a reproduction curve of a single-species population is in large part formed in accordance with the varying combinations of these components and elements: sociality and competition as two major components, the latter comprising the two sub-elements (scramble and contest). Nonetheless, we would not see a sharp distinction among these components or elements in terms of the variation in the reproduction curves in Figure 6.2 because the variation is a continuum. We may see a difference only between the two curves sufficiently apart from each other

140 · Scramble and Contest Competition

in terms of the value of h, e.g. between the two extremes: the Malthusian progression ($h = -1$, or no competition) on one end and, on the other end, the curve with the steepest downward slope ($h \to \infty$) that is largely made up of the scramble element. Consequently, it would be unlikely that we see any example that is made up of a single component or element. In other words, among the reproduction curves in Figure 6.4, we cannot single out one category to assert: this represents sociality and that represents competition. By the same token, an attempt to single out a category that represents either element (scramble or contest) of competition would certainly fail.

6.8 Concluding Remarks of Chapter

The distinction between scramble and contest competition serves as a useful concept for analysing and interpreting a reproduction curve as we observe. It also aids us to understand the evolutionary process of such things like territoriality and asynchronous hatching as we see ubiquitously among avian species. In this regard, compared with the scramble, the contest is likely a more prevalent element and an advanced form of competition. However, the distinction is subtle, so subtle in fact that it would be impossible to find a typical example of either element in a pure form. In this sense, the notion of the two types of competition, as verbally stated in the first paragraph of the present chapter should be taken as a sort of allegory: a literal interpretation of it may mislead. Nonetheless, the allegory succinctly depicts the nature of the two elements of competition in an intuitively comprehensible manner: it only needs to be amended in a classroom.

Appendix 6A: The Logarithm of a Negative Real Number is a Complex Number

Some time ago, a colleague of mine asked: 'Tom, I know the square root of a negative number is an imaginary number. But, I do not think the logarithm of a negative number exists, does it?' Yes, it does, but it is a complex number.

Let z represent a complex number, written commonly as $z = x + iy$, in which both x and y are real numbers and i, the imaginary unit, defined as the square root of -1, i.e. $i = \sqrt{(-1)}$ or $i^2 = -1$. Geometrically, z is represented in the complex plane with the real and imaginary axes: see

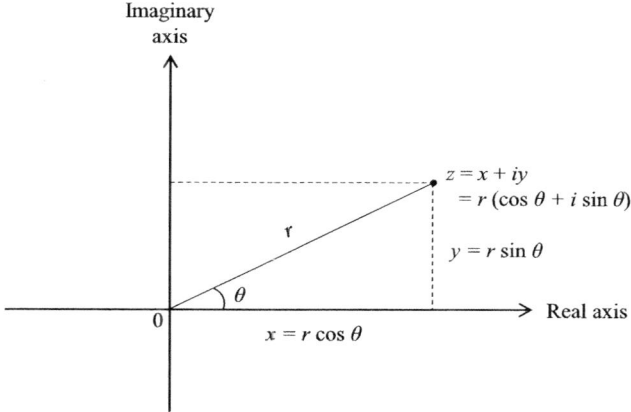

Figure 6A.1 Graphical representation of a complex number.

Figure 6A.1. It is easy to see that z can also be written, using the polar forms: $x = r \cos \theta$ and $y = r \sin \theta$ where r is the distance (always non-negative and real in polar form) between the origin and the point z, and θ in radians (see Note below), i.e.

$$z = r(\cos \theta + i \sin \theta). \tag{6A.1}$$

Now comes the famous (in mathematics) but really weird formula, called Euler's formula, named after the great eighteenth-century Swiss mathematician Leonhard Euler, i.e.

$$\cos \theta + i \sin \theta = e^{i\theta} \tag{6A.2}$$

[Note: How did he find the formula? I will show you in a moment. By the way, the constant $e = 2.71828...$ is commonly known as the base of natural logarithms, but mathematicians call it Euler's number e because it was Euler who assigned this particular letter to the constant.]

Substituting the relationship (6A.2) for the same in (6A.1), he found:

$$z = re^{i\theta}. \tag{6A.3}$$

Substituting $\theta = \pi$ in (6A.1), we find (he found) $z = -r$, because $\cos \pi = -1$ and $\sin \pi = 0$. That is, $z = e^{i\pi} = -r$, a negative real number. But, $z = -r$ holds also for $\theta = (1 \pm 2n)\pi$, $n = 0, 1, 2, 3, \ldots$, because the trigonometric functions have the cycle of the period 2π.

Now, take the logarithms on both sides of (6A.3) with $z = -r$ and $\theta = (1 \pm 2n)\pi$, we find the logarithm of $z = -r$ (a negative real number) to be:

$$\ln(-r) = \ln(r) + i(1 \pm 2n)\pi, n = 0, 1, 2, 3, \ldots$$

We see that the logarithm of a negative real number is a multi-valued complex number with the cycle of period 2π. Normally, it is sufficient to consider the principal value at $n = 0$, which is written, using the uppercase letter L in Ln, i.e.

$$\text{Ln}(-r) = \ln(r) + i\pi. \tag{6A.4}$$

Although an imaginary number (as the square root of a negative real number) is taught in school mathematics, few school textbooks talk about the logarithm of a negative real number. Why? Perhaps, I think, because it is just another complex number, and unlike the imaginary unit i, it offers no more unique mathematical utility.

Euler's formula in (6A.2). You may be curious about how Euler had this formula. It is quite straightforward, once you know how he did it. So, here we go.

In Chapter 3, I introduced the Maclaurin series. To repeat it with a bit of notational changes from the generic variable u to the angle θ in radian:

$$f(\theta) = f(0) + f'(0)\theta + f''(0)\theta^2/2! + \ldots + f^{(j)}(0)\theta^j/j! + \ldots \tag{3.8 rpt}$$

where $f^{(j)}(0)$ designates the j-th derivative $f^{(j)}(\theta)$ at $\theta = 0$ and is a constant, given $j = 0, 1, 2, 3, \ldots$ Now, consider that $f(\theta)$ is the exponential function e^θ, i.e. $f(\theta) = e^\theta$, the derivative of which is itself, i.e. $f'(\theta) = e^\theta$, and so are all higher-order derivatives $f^{(j)}(\theta) = e^\theta$ for all $j = 1, 2, 3, \ldots$

Then, $f^{(j)}(0) = e^0 = 1$ for all j. Substituting 1 for every one of the $f^{(j)}$ on the right-hand side of equation (3.8 rpt), and replacing θ with $i\theta$ (where i is the imaginary unit), we find:

$$e^{i\theta} = 1 + i\theta + (i\theta)^2/2! + \ldots + (i\theta)^j/j! + \ldots$$

Now notice that the expression i^j ($j = 0, 1, 2, 3, \ldots$) yields the series of four different values that are repeated in cycles as below:

j	0	1	2	3	4	5	6	7	8	9	10	11...
i^j	1	i	-1	$-i$	1	i	-1	$-i$	1	i	-1	$-i$...

So, we find:

$$e^{i\theta} = 1 + i\theta - \theta^2/2! - i\theta^3/3! + \theta^4/4! + i\theta^5/5! - \theta^6/6! - i\theta^7/7! \\ + \theta^8/8 + i\theta^9/9! - \theta^{10}/10! - \ldots$$

Meanwhile, using the Maclaurin series (3.8 rpt), we can expand the trigonometric functions on the left-hand side of (6A.2):

$$\cos\theta = 1 - \theta^2/2! + \theta^4/4! - \theta^6/6! + \theta^8/8! - \theta^{10}/10! + \ldots$$

$$i\sin\theta = i\theta - i\theta^3/3! + i\theta^5/5! - i\theta^7/7! + i\theta^9/9! - \ldots$$

We see then that the sum ($\cos\theta + i\sin\theta$) is equal to $e^{i\theta}$.

Appendix 6B: How to Estimate Parameters (R_m, h, c/h) to Fit Model (6.1a) to the Observed Reproduction Curve in Figure 6.3a

Unlike a linear regression, there is no standard way of fitting a non-linear model, like (6.1a) in the main text, to an observed curve: how to do it depends on the model. It can be done approximately by iteration (repetitive numerical computations), starting with a set of guesstimated values, until the result converges to a desired degree. The following is my initial guestimate.

First, recall Figure 6.1 (or 6.2) in the main text and remember that R_m is the upper asymptote of the reproduction curve. In the meantime, notice that the observed curve in Figure 6.3a is quite close to the asymptote. In particular, the highest observed value (the first circle on the left-end of the observed curve) is 32.6. Hence, $\log(32.6) \approx 1.513$ is my initial guessed value for R_m.

The value of h is computed as follows. Recall that the slope ($\tan\theta$) of a reproduction curve (6.1a) at X^* is given by the derivative in (6.2a). Thus:

$$\tan\theta = dR_t/dX_t|_{X^*}$$
$$= -(h+1)\{\text{antilog}[R_m/(h+1)] - 1\}/\{\text{antilog}[R_m/(h+1)]\}.$$
(6.2a rpt)

In the observed reproduction curve (the solid circles in Figure 6.3a), the slope at the equilibrium point X^* (angle θ in degrees) is about $-57°$. Substituting the guessed value $R_m \approx 1.513$ and the observed value $\theta \approx -57°$ (or $\tan\theta \approx -1.54$) in the above equation, and solving for $(h+1)$, we find $h \approx 0.8$. [Note: An algebraic solution for $(h+1)$ in the above equation is difficult to obtain. But, reversing the calculations, i.e. substituting the guessed value 0.8 on the right-hand side, we find $\tan\theta \approx -1.540153\ldots$ Good enough!]

Scramble and Contest Competition

As for (c/h), recall that $R_t = 0$ at the equilibrium point x^*, and hence we find by model (6.1a convert) that:

$$0 = R_m - (h+1)\{\log[1 + (c/h)x^*]\}$$

and solving for (c/h) we have:

$$(c/h) = \{\text{antilog}[R_m/(h+1)] - 1\}/x^*.$$

Meanwhile, in Figure 6.3a, we can graphically read an approximate value of x^*: it turns out to be about 610. Substituting this approximate value, and those for $R_m/(h+1)$ already guesstimated, we find $(c/h) \approx 0.01$. Altogether, the set of my initial guesses for $(R_m, h, c/h)$ is (1.513, 0.8, 0.01). After small adjustments by iteration (trial and error, in fact), I found $R_m = 1.52$ and the final set of estimates $(R_m, h, c/h) = (1.52, 0.8, 0.01)$ which yielded a near-perfect fit.

7 · *Regulation of Populations: Its Myths and Real Nature*

7.1 A Brief History

During the 1950s, my apprenticeship years, hot debates were ongoing in animal ecology as to what factors control or regulate animal populations. Prominent ecologists in the world across Europe, North America, and Down Under (the hot bed, in fact) were divided into two schools of thought: one emphasizing the essential roles played by biotic factors, the other stressing the importance of physical factors, like climatic influences. However, the origin of the debates goes way back to the early 1900s, when two American entomologists, L. O. Howard and W. F. Fiske, recognized two distinct categories of mortality factors, 'facultative' and 'catastrophic', in their 1911 article. A couple of decades later, H. S. Smith (1935), a colleague of Howard's, coined the terms that are now universally used in ecology: 'density-dependent' and 'density-independent' factors.

The focus of the Great Debates was to answer the question: 'How are the populations of organisms so controlled or regulated as to persist?' In both schools, it was intuitively conceived that, for a given population to persist for a sufficiently long time, there must be a mechanism that would, on the one hand, discourage a high population to increase further and would, on the other, prevent a low population from continuing to decrease to extinction. From this intuitive perception of population regulation stemmed the notion that the action of the regulatory mechanism should necessarily depend on population density. This intuitive notion was the basis of the contention for the biotic-factor school. In contrast, the climatic-control school emphasized the importance of environmental (density-independent, *sensu* H. S. Smith) factors on the basis of their perception that climatic influences appeared to be dominant factors in determining many insect population processes.

The theory of population regulation by biotic factors was championed by the Irish-Australian entomologist A. J. Nicholson, strongly supported by my ex-supervisor David Lack (1954), together with Charles Elton and George Varley, of the Department of Zoological Field Studies, Oxford, UK. In the climatic (abiotic) factor school, the idea of the German-Israeli entomologist F. S. Bodenheimer inspired the two Australian ecologists H. G. Andrewartha and L. C. Birch (1954), who strongly advocated the climatic control theory in their well-known textbook.

Carefully studying the arguments made by both schools, I tended to be inclined to the biotic theory because its contention made more logical sense to me. Nonetheless, I was not entirely sure about it. Rather, I had a feeling that something was missing in both schools in their contentions. Moreover, it seemed to me that many of the arguments were based largely on empirical facts, or even on myths, rather than on logically and rigorously induced principles.

It was not until I went to study stochastic processes under Professor Patrick A. P. Moran in Australian National University in the mid-1970s when I began to understand what was missing in the great (but indecisive) debates: neither school had a clear idea of what they meant by 'persistence', 'regulation', or 'control'. Perhaps, each person thought he (there were few 'she' participants in the debates as far as I remember) had a clear idea, but I found no one had defined these terms in a rigorous manner. Without an appropriate definition, how could one argue about what 'regulates' or 'controls' populations so that they 'persist'?

Here, I attempt first to find an appropriate way to define the concept of 'population persistence'. Then, I show that directly from the definition stem the criteria for assessing the roles played by 'density-dependent' and 'density-independent' factors in 'regulating' or 'controlling' populations. So, to delve into this issue, I should begin with some basics of population dynamics as stochastic processes.

7.2 Biological Population Processes As Stochastic Processes

A stochastic process is, in short, a set of random events that occurs sequentially in time (or in space). The analyses of these processes, as sequences, are distinguished from the usual (or sometimes referred to as 'classical') statistical analyses of events that do not occur in sequence (or need not be treated as sequential events, even if they were), e.g. some attributes (height, weight, etc.) of the collection of all male babies born in Tokyo in 1930 (of which I was an element) at a given age-class.

In this classical case, we may characterize the collection (population) of these babies in terms of the (frequency) distribution of, say, their weights,

7.2 Biological Population Processes · 147

and the distribution is in turn characterized by a set of moments, most importantly the mean (the first moment) and the variance (the second moment). Although it is impractical to weigh every one of the babies at the same time, we know (or expect) that these moments exist conceptually as the 'population mean' and 'population variance'. [The word 'population' here means the conceptual existence of all individuals (elements) concerned that we do not, or need not, actually see.] So, letting w be the weight of each baby, we write the first moment (population mean) and the second moment (population variance) of the w as $E(w)$ and $E[w - E(w)]^2$, respectively: the expression 'E(something)' reads 'the expectation of (something)'. Because the values of these expectations (or expected values) are usually unknown (or often impractical to know), we estimate them by sampling as a 'sample mean' and a 'sample variance'.

In contrast to the above classical example, we can think of changes in weight of a baby from its birth to adulthood and beyond as a stochastic process. To statistically characterize the process, consider again the population of all male babies born in Tokyo in 1930. As their weight changes with age $(0, 1, \ldots, i, \ldots)$, we can think of the mean weight of all boys in the population at age i, which defines the expectation $E(w_i)$. Furthermore, as a stochastic process, we also consider the generalized second moment, known as the 'covariance', between the weights of the boys at ages i and j. The covariance is formally defined as the deviation of w_i from its mean $E(w_i)$, i.e. $[w_i - E(w_i)]$, multiplied by the same with respect to the age-class j. That is, the expectation of the product $[w_i - E(w_i)][w_j - E(w_j)]$, i.e. $E\{[w_i - E(w_i)][w_j - E(w_j)]\}$, is used as a measure of the temporal changes in mean weight. This is called an 'autocovariance' because it is the covariance between the same attribute (weight) of the boys at different points in time (ages): the variance in a given age-class, say $E[w_i - E(w_i)]^2$, is a particular case of the covariance in which $i = j$. [Note: Inasmuch as the expectation $E(w_i)$ can be estimated by averaging the weights of sampled individual boys at age i, written $M(w_i)$, the expectation $E\{[w_i - E(w_i)][w_j - E(w_j)]\}$ can be estimated by averaging the observed products $[w_i - M(w_i)][w_j - M(w_j)]$ among the sampled boys at ages i and j: for further technical details, see a textbook of statistics.]

Likewise, the temporal changes in the size of an animal population as a stochastic process, say $\{x_t\}$ at time (or generation) t as well as its rate of change $r_t = x_{t+1}/x_t$, would be characterized by these expectations: the means and autocovariances. Well, that is easier said than done, in fact: there is a big pragmatic problem with studying natural populations. We

148 · **Regulation of Populations**

may in most cases have just a few, often only one, series of observations in many field studies. A single series of population changes is compared to the sequential observations of weight changes in a single boy: how can we know these expected values of the population of boys by observing just one individual? I shall talk about the nature of this problem later when I deal with a practical procedure of estimation. For now, let us assume that we generate a number of series, using an idealized model process, to study the nature of these mathematical expectations (or simply expectations) with the aim of defining the concept of population persistence.

7.2.1 Expectations in a Model Population Process

To conceive the idea of 'expectations' in principle without complication, consider that we generate a population series $\{X_t = \ln(x_t)\}$, using the (simple) stochastic process model (4.9) in Chapter 4, repeated below:

$$R_t = (R_m + \varepsilon_t) - \exp[(C + \zeta_t) + X_t] \qquad (4.9 \text{ rpt})$$

in which: $R_t = \ln(r_t) = \ln(x_{t+1}/x_t) = X_{t+1} - X_t$; $R_m = \ln(r_m$: mean potential reproductive rate of individuals); $C = \ln$(intensity of within-population competition); ε_t and ζ_t represent exogenous influences. [Note: I may call R_t and X_t simply 'rate of change' and 'density', although these are $\ln(r_t)$ and $\ln(x_t)$.] For further simplicity, let us consider that only parameter R_m is subjected to the exogenous influences $\{\varepsilon_t\}$. Thus, (4.9) is simplified to:

$$R_t = (R_m + \varepsilon_t) - \exp(C + X_t). \qquad (4.9 \text{ simp})$$

Furthermore, let us assume that ε_t is a random variable. Then, R_t becomes a random variable. Accordingly, X_{t+1} ($= R_t + X_t$) become random variables as well. Altogether, starting from an arbitrarily given X_0, the series $\{X_t; t = 1, 2, 3, \ldots\}$ becomes a stochastic (random) series.

Now, to generate the stochastic series $\{X_t\}$, we must first generate (using a computer-based random number generator) a series of random numbers $\{\varepsilon_t\}$. For this purpose, I consider the very basic set of random numbers, called "independent, identically distributed" (or iid) random numbers as explained below.

7.2.2 Independent, Identically Distributed (IID) Random Numbers

Consider a series of random numbers $\{\varepsilon_i\}$ in which i may vary like: 1, 2, …, t, or 0, 1, 2, …, $(t-1)$, whichever suits you. If the two elements that are apart from each other by any number of time steps (e.g. ε_i and ε_j, $i \neq j$) are uncorrelated with each other, they are called 'independent' random numbers. That is, the autocovariances among all elements of the series $\{\varepsilon_i\}$ are identically zero, the feature that is standard in computerized random-number generators.

Now, as random numbers, they are characterized by their distribution, e.g. normally, uniformly, etc., distributed. Furthermore, these distributions are quantitatively characterized by their moments, the most important ones being the mean (the first moment) and the variance (the second moment). If all elements of the series $\{\varepsilon_i\}$ are distributed in the same manner (e.g. normally distributed), sharing the same mean and variance (in value), these elements are said to be 'identically distributed'.

Altogether, the series $\{\varepsilon_i\}$, if made up of the 'independent' and 'identically distributed' elements, it is called an iid random series. [Note: Incidentally, an iid random number (in particular, normally distributed) is often referred to by the jargon 'white noise' in physics, implying that it is an unwanted noise that may obscure the 'signal' that is wanted. I would avoid that jargon because, in population ecology, exogenous influences are in general important signals: it may become a 'noise' only when one wants to know about endogenous features of the process under study.]

An iid series of random numbers can be readily generated by computer. To do it, you specify (tell the computer) the distribution type, as well as the values of the mean and the variance you want: in the present book, I usually choose the normal (but occasionally the uniform) distribution, depending on its intended use.

7.2.3 Stochastic Population Processes

Now, consider that we generate a number of iid (normally distributed with the 0 mean and an arbitrary variance) series $\{\varepsilon_t\}$ and incorporate each one into the foregoing model (4.9 simp) to generate so many number of series $\{X_t\}$. Then, each of the $\{X_t\}$ series is compared to the series of the weight records of a single baby boy/girl. [Remember, however, that $X_t = \ln(x_t)$.] Thus, we can think of the 'population (collection or ensemble)' of these model series with their expectations, namely population mean $E(X_i)$ and population autocovariances $E\{[X_i - E(X_i)][X_j - E(X_j)]\}$ in which, as already mentioned, the variance $E[X_i - E(X_i)]^2$ is a special case for $i = j$.

I now show that the notion of population persistence can be defined in terms of these expectations.

7.3 Defining Population Persistence

Perhaps an intuitive idea of population persistence is that the temporal path (graph) of the series $\{X_t\}$ is, although fluctuating continually, somehow confined within a certain range (say A) at a given level, say M. This attribute may be mathematically expressed in terms of the absolute value of the difference $(X_t - M)$ to be no greater than A, i.e. $|X_t - M| \leq A$. Pragmatically, however, this definition is not useful because it does not provide information on how the series $\{X_t\}$ can comply with the constraint.

Furthermore, we must recognize that nothing is perpetual in the whole universe, and it is pointless to think about a population perpetually staying within a given range in an absolute sense. Besides, practically, we can observe only a part of the whole length of the process, which usually is not long enough to be certain that the population would continue to stay within the range. So, from a pragmatic point of view, I suggest that we abandon the notion of persistence in the absolute sense. But, what is an alternative?

7.3.1 Second Moment as a Measure of the Likelihood of Population Persistence

A pragmatic alternative is to replace the expression $|X_t - M|$ with the squared deviation of X_t about M, i.e. $(X_t - M)^2$. In other words, if the series $\{X_t\}$ is confined mostly within a certain range for a sufficiently long period of time, then the second moment of the series $\{X_t\}$ about M, i.e. $\mathrm{E}(X_t - M)^2$ as a statistical measure of dispersion of X_t about M, should be finite-valued, say A. Thus, we define the persistence of a population to be:

$$\mathrm{E}(X_t - M)^2 = A < \infty. \tag{7.1}$$

[I use the expression '$A < \infty$' as equivalent to symbolically saying: 'A is finite'. From now on, A may be dropped for simplicity.]

An important point to keep in mind is that definition (7.1) would not mean that the series $\{X_t\}$ stays within a fixed range all the time in the absolute sense: the series may stay within a certain range most (say, 95%) of the time, but may, every now and then, come outside with no specific limit in the absolute sense. Thus, we may interpret the value of A in (7.1) as an ecological measure of the likelihood of the population staying within a certain range about M: for instance, the smaller the value of A, the greater the likelihood, or vice versa.

Altogether, I suggest that we think of the persistence of a population in terms of the likelihood of the population under observation to comply with (7.1) as the statistical requirement or stipulation. If the observed series complies, we may say that the series is *in the state of persistence* or in a persistent state, even if it might someday go extinct by accident, depending on the variance of the exogenous influences $\{\varepsilon_t\}$. So, let us look into the mechanisms required for a population to be in a persistent state.

7.4 Investigations into Mechanisms for Persistent State

Stipulation (7.1) is a good way of defining a measure of the likelihood of the state of persistence. As it stands, however, expression $E(X_t - M)^2$ per se is not of an insightful form. This is because the population size X_t is a consequence of the rate of change in population, given the size in the preceding generation X_{t-1}. So, rather than look into the consequence, we should look directly into the cause, i.e. the rate of change. In other words, we should express the left-hand side of (7.1) in terms of the rate of change. This can be readily done as shown below.

Recall the definition of the net (per-capita) rate of change r_i in population from x_i to x_{i+1}, i.e. $r_i = x_{i+1}/x_i$, which, after an ln- (or log-) transformation, is written $R_i = X_{i+1} - X_i$, or transposed: $X_{i+1} = R_i + X_i$. Then, starting with an arbitrary initial size, say X_0, we have the following rows of relationships:

$$X_1 = R_0 + X_0$$
$$X_2 = R_1 + X_1$$
$$\dots\dots\dots\dots$$
$$X_{t-1} = R_{t-2} + X_{t-2}$$
$$X_t = R_{t-1} + X_{t-1}.$$

Summing all rows on both sides of the equality signs, we find that the sum ΣX_i over $i = 1$ to $t-1$ on the left-hand side is cancelled by the same sum on the right. That is, what remains is X_t on the left, and on the right X_0 plus the sum of R_i over $i = 0, 1, 2, \ldots, (t-1)$, written ΣR_i. That is, we find:

$$X_t = \Sigma R_i + X_0, (i = 0, 1, 2, \ldots, t-1). \tag{7.2a}$$

Transposing X_0 to the left-hand side, then squaring both sides and taking expectations, we have:

$$E(X_t - X_0)^2 = E(\Sigma R_i)^2. \tag{7.2b}$$

Thus, setting $X_0 = M$ conveniently, we see that, for the population to meet stipulation (7.1) to be in a persisting state, it is required that the expectation $E(\Sigma R_i)^2$ be bounded, i.e. finite, written compactly as:

$$E(\Sigma R_i)^2 < \infty. \qquad (7.3)$$

I now attempt to interpret stipulation (7.3) in terms of the concepts of density dependence and independence. Let us look into density dependence first.

7.4.1 Concept of Density Dependence in the Context of Population Persistence

In Figure 4.3, and its variants Figures 4.5 and 4.7 in Chapter 4, I showed how to generate population series by means of a reproduction curve (a plot of reproductive rate R_t against population density X_t), and some typical examples of the series generated thus were shown in Figure 4.6. We see that all these series converged to a bounded range of variations. This implies their compliance with stipulation (7.3). [Note: The series in Figure 4.6 are not exposed to any exogenous influences. So, expectation (7.3) is interpreted as equivalent to the rate of change in endogenous form to be bounded, i.e. $(\Sigma R_i)^2 < \infty$.] From this point of view, I suggest that we look into the reproduction curves in Figure 6.2 (extended versions of the aforementioned curves in Chapter 4) to recognize the following two endogenous attributes.

Attribute (i). All (except one on the top) reproduction curves in Figure 6.2 decrease as X_t increases but are bounded from above by the maximum reproductive rate R_m. Thus, the (intergeneration reproductive) rate R_t is density-dependent in that it is a function of density but in a specific manner (or in a narrow sense): it is a monotonically decreasing function of X_t. [Note: In a broader sense, R_t is 'density-dependent' if it is a function of X_t in any manner: decreasing, increasing, curvilinear, linear, or a combination of all these. However, in the context of the persistent state of populations, I use the term 'density-dependent' in the narrow sense: a monotonically decreasing function of X_t, as in Figure 6.2.]

Attribute (ii). The parameter R_m is positive and finite such that all curves (in Figure 6.2) intersect the horizontal ($R_t = 0$)-axis from above to ensure the existence of an equilibrium density, designated X^*. [Note: As shown in Section 4.9, Chapter 4, there are two categories of X^*. In one, X^* is stable such that every one of the generated series $\{X_t\}$ converges to it: this occurs if the slope of the reproduction curve is no steeper than

$-45°$ at X^*. In the other category, in which the slope is steeper than $-45°$, X^* becomes unstable such that the series $\{X_t\}$ diverges from X^* in its very vicinity. Nonetheless, every one of the series generated in this category converges to a certain (bounded) area around X^* (e.g. a limit cycle): this attribute is ensured because the reproduction curves are all bounded from above with the maximum R_m.]

Altogether, we see that a population can be in a persistent state if its reproduction curve is density-dependent, *sensu* attribute (i). However, it is important to notice that a density-dependent reproduction curve does not necessarily ensure a persistent state of the population. Quite clearly, if R_m is negative, that is, lacking the attribute (ii), the whole section of the curve would be below the $(R_t = 0)$-axis, meaning that the population goes consistently to extinction. Yet, the reproduction curve remains as density-dependent in the narrow sense. Conversely, what happens if R_m has no upper bound? The amplitudes of population oscillation may increase without bound, i.e. stipulation (7.3) may not be met. Yet, again, the reproduction curve remains density-dependent.

Altogether, the density dependence of a reproduction curve per se would be unable to maintain the population in a state of persistence because it is incapable of controlling R_m to be within an appropriate (positive and finite) range. However, the foregoing investigations of density dependence have been based on a reproduction curve in its endogenous form. So, in order to fully understand the nature of density-dependent processes, we should investigate the effect of exogenous influences on them.

7.5 Density-Dependent Processes under Exogenous Influences

Consider a situation in which the endogenous parameter R_m is subject to a fluctuating environment, i.e. represented by the random number ε_i. To investigate what we would see, let us incorporate ε_i into model (6.1a) of Chapter 6 that has generated all reproduction curves in Figure 6.2. For simplicity again, I consider a situation in which only R_m is subjected to ε_i such that the resultant (endo-exogenous) process model is given by:

$$R_i = (R_m + \varepsilon_i) - (h+1)\{\ln[1 + (c/h)\exp(X_i)]\}.$$

Also, assume a simple situation in which ε_i is an independent, identically distributed (iid) random number with mean μ_ε and variance σ_ε^2 for all

$i = 0, 1, 2, \ldots, t-1$. Thus, R_i is a random variable whose expectation is given by:

$$E(R_i) = (R_m + \mu_\varepsilon) - (h+1)E\{\ln[1 + (c/h)\exp(X_i)]\}.$$

On the surface of this form of the model, it appears that the effect of ε_i merely changes the value of R_m as the sum $(R_m + \mu_\varepsilon)$. If so, nothing is new here: for the (model) population to be in a state of persistence, it is necessary that the sum $(R_m + \mu_\varepsilon)$ is positive and finite, i.e. equivalent to the aforementioned endogenous attribute (ii). However, a little more careful consideration is necessary in general situations.

7.5.1 Likelihood of Persisting State in the Fluctuating Environment

Now that ε_i is a random number, we cannot be absolutely sure that a positive value of the expectation $(R_m + \mu_\varepsilon)$ ensures the persistent state of the population. It depends on the magnitude of the variance σ_ε^2: a large σ_ε^2 might make the value of $(R_m + \varepsilon_i)$ too low (even negative), or too high, too often. In other words, under a fluctuating environment, the likelihood of a persistence state depends on the magnitude of the variance σ_ε^2: the smaller the variance the higher the likelihood, and vice versa.

Furthermore, we should consider a situation in which μ_ε changes with time, e.g. an effect of climate changes, such that the sum $(R_m + \varepsilon_i)$ may have a trend: in other words, $\{\varepsilon_i\}$ is no longer an iid series and may influence the dynamical pattern of the population series $\{X_t\}$ in violation of stipulation (7.3). Exogenous influences may act on parameter c, too, by laterally shifting the equilibrium density X^* of the reproduction curve, even though the curve remains density-dependent in the narrow sense.

Altogether, density dependence would not always ensure the persistent state of a population under exogenous influences. This is a crucial issue in the investigation into the state of population persistence. However, before we proceed to delve into this issue, I suggest that we look into the nature of density-independent processes as a preliminary step.

7.6 Density-Independent Processes

7.6.1 Graphical Perception

Among all the reproduction curves in Figure 6.2, the one on the very top is the odd one out in that it is a straight horizontal line in which

7.6 Density-Independent Processes · 155

$R_t = R_m = 2$, a positive constant regardless of density X_t. Hence we may say that it is a density-independent process. This horizontal line, as already explained, is a reproduction 'curve' of the Malthusian process that increases in a geometric rate without bound. What if R_m was negative? The horizontal reproduction line is below the ($R_t = 0$)-axis, and it would generate a reversal of the Malthusian, i.e. an ever-decreasing population without bound (i.e. $X_t \to -\infty$), leading consistently to its inevitable extinction. Either way, the population cannot be in a persistent state.

But what if $R_t = R_m = 0$? The sum ΣR_i over $i = 0, 1, 2, \ldots, t-1$ (in endogenous form) would be identically 0. Then, from the exact relationship (7.2a), we see $X_t = \Sigma R_{t-1} + X_0 = X_0$. This means that the population would deterministically stay at the initial position X_0 forever. It appears that the population is in a state of equilibrium forever. Indeed, this attribute meets stipulation (7.3) if the effect of exogenous influences is nil. But $R_i = 0$ identically for all i means that the rate of change in population is density-independent. Then, it might sound logical to think that, if a certain condition is met, a density-independent process could maintain the population in an equilibrium state for an arbitrarily long period of time. Right? Not quite!

In applied mathematics, this state of equilibrium is said to be 'neutral' in that the population would stay wherever it was placed, like a golf ball sitting on a flat green: it will stay wherever it is initially placed, provided that there is no strong wind. Thus, the population would remain in a persistent state as long as no exogenous influence perturbed this equilibrium state. In reality, this would not happen because the rate of change R_i is always under the influence of a fluctuating environment. It need not even be an extreme situation (like the nasty blowing wind in some British Open golf tourneys) to disturb the neutral state; just an extremely light perturbation will do. What could happen then is the following. [Note: In reality, the probability of R_m being equal exactly to 0 would be nil. However, what I describe below approximately applies when R_m is close to 0.]

7.6.2 Density-Independent ($R_t = 0$) Process under Exogenous Influences

Starting at the initial position X_0, the rate of change R_0 in general would not be zero, but would be likely displaced as much as ε_0 such that $X_1 = \varepsilon_0 + X_0$. Similarly, the following $X_2 = \varepsilon_1 + X_1$ which is equal to $\varepsilon_0 + (\varepsilon_1 + X_0)$, and so on. Then, by relationship (7.2a) in which $R_i = \varepsilon_i$

for all i, we find $X_t = (\Sigma \varepsilon_i) + X_0$, $i = 0, 1, 2, \ldots, (t-1)$. We see that, in general, X_t would not be equal to the initial state X_0. Instead, its fate would depend solely on the sum $(\Sigma \varepsilon_i)$ in which i varies from 0 to $(t-1)$.

Let us assume, for simplicity, that ε_i is an iid random number, normally distributed with mean 0 and variance of an arbitrary value. Then, you might have thought that, because the series $\{\varepsilon_i\}$ fluctuates about 0 with an equal likelihood of deviating on either side of 0, a deviation on one side would cancel one on the opposite side. As a result, you might have thought that their sum would stay close to zero most of the time. Right? Wrong! What in fact happens is that these deviations, i.e. $\{\varepsilon_i\}$ from 0 by chance (or 'errors', if you like), tend to accumulate in their sum over time rather than cancel each other. Consequently, the sum would not stay near 0, but would wander around in an uncontrolled (unpredictable) manner. In applied maths, this tendency is known as a 'random walk'.

A typical example of a random walk is the well-known Brownian motion in physics. Consider a pinch of pollen grains floating on water. Were the surface perfectly still, each grain might have stayed there, but this does not happen because the grains are tossed around by the random movements of the water molecules (H_2O). As a result, at a given moment, the position of a given grain relative to its initial position is unpredictable: after a while, some grains might stay nearby their initial positions, but others might stray afar. That is, the path of a grain exhibits no particular tendency but random walk. [Note: The random walk is often and aptly referred to as the 'drunkard's walk', as some of you might have personally experienced.] The same applies to a population process in principle, but the representation of its motion differs slightly.

Notice that the aforementioned example of Brownian motion is the continuous-time motion of a grain on a two-dimensional surface with a hidden time axis. Compared with this, the movement of the population process $\{X_t\}$ is usually represented by a series of vertical deviations from the horizontal time axis through the origin X_0. That is, the random walk in a population exhibits an uncontrolled wandering path along and about the time axis. Let me show you some examples by simulation.

7.6.3 Examples of Random Walk Paths in Population

Using a computer-based random number generator, I generated five series of iid random numbers $\{\varepsilon_t\}$ to represent the series $\{R_t\}$, i.e. $R_t = \varepsilon_t$, specified as: normally distributed with 0 mean and (arbitrarily chosen) 0.25 variance. Figure 7.1a shows five sections of the iid series in

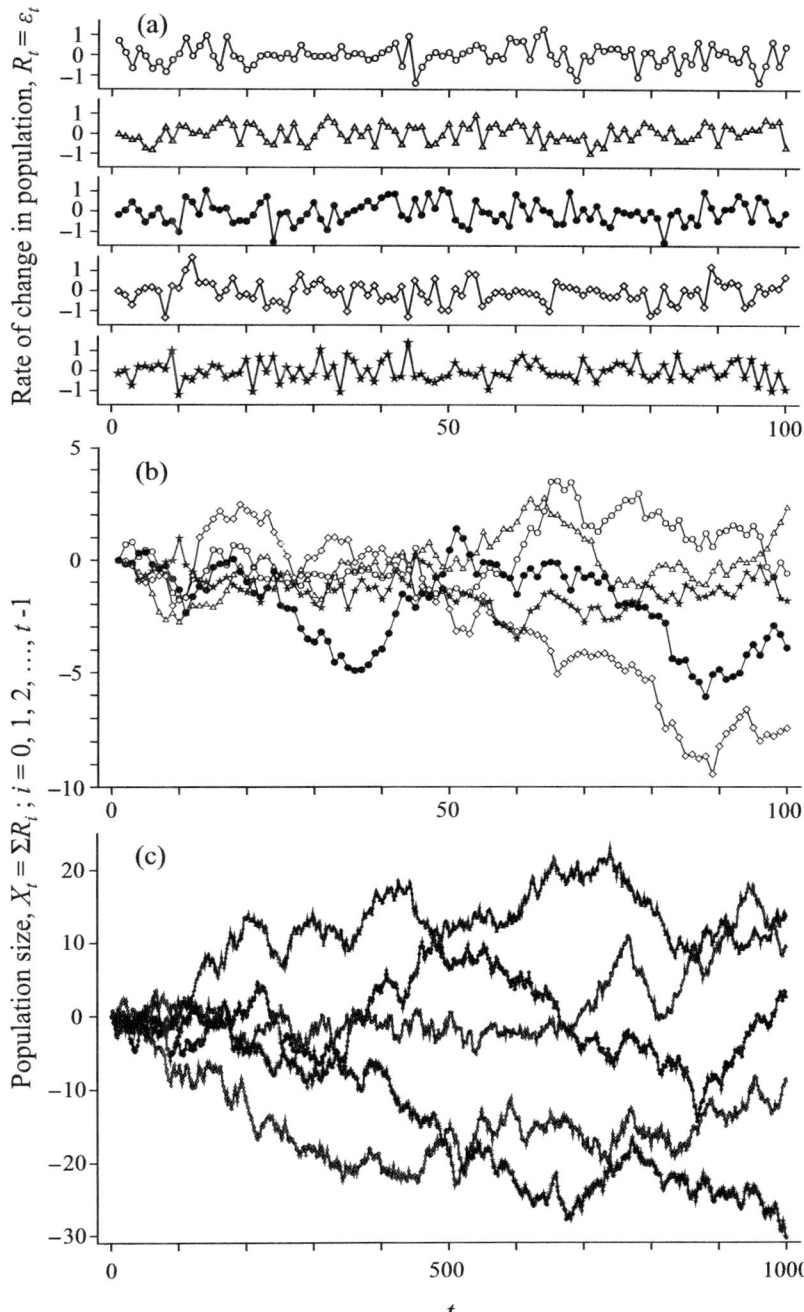

Figure 7.1 Examples of random walk paths. (a) A series of random numbers. (b, c) Random walk paths as the sums of the random number series in (a).

length 100. Then, I calculated the sum ΣR_i (over the entire length) in each series and, incorporating these sums in relationship (7.2a), I generated five series $\{X_t\}$ of length $t = 100$, all starting with $X_0 = 0$ for ease of comparison. These are plotted against t in Figure 7.1b. We see each series exhibits no fixed pattern: each one wanders around every which way; some appear to be increasing; some decreasing; some in between; or a combination of all these.

In Figure 7.1c, I extended these series $\{X_t\}$ further to $t = 1000$. Again, we see no fixed pattern in each individual series, but all of them collectively (or as an ensemble) appear to exhibit certain tendencies in two aspects: (a) the ensemble appears to arrange itself (sort of) symmetrically about the $(X = 0)$-level; (b) the range of deviations about the 0-level (i.e. the variance) of the ensemble appears to increase (fanning out) as time goes on with no sign of convergence to a finite value.

Altogether, a density-independent population process in a neutral equilibrium state performs a random walk under (uncorrelated) exogenous influences and does not comply with stipulation (7.3): that is, a random walk path is in a non-persistent state. [Note: Strictly speaking, a density-independent population process would perform a random walk when the exogenous influences are uncorrelated random numbers. In general, those influences can be correlated, and the resultant population series may (mathematically) conform to stipulation (7.3), as I will show in due course.]

The foregoing examples of density-dependent and -independent processes provide intuitively comprehensible bases for understanding the nature of the persistent (or non-persistent) state of populations. However, the following algebraic analyses of (7.3), albeit a little tedious at times, would provide a much deeper insight. So, here we go.

7.7 Algebra of Stipulation (7.3) for Population Persistence

Consider a general case of the stochastic series (process) $\{R_i\}$ in which the mean $E(R_i)$ and the variance $E[R_i - E(R_i)]^2$ differ, respectively, from $E(R_j)$ and $E[R_j - E(R_j)]^2$ for $i \neq j$, but R_i is correlated with R_j. Under these assumptions, let us do some manipulations with the sum ΣR_i in (7.3). To begin, I create the artificial term $[-\Sigma E(R_i) + \Sigma E(R_i)] = 0$. Then, adding this 0-term to ΣR_i and rearranging, we have:

$$\Sigma R_i = \Sigma R_i - \Sigma E(R_i) + \Sigma E(R_i)$$
$$= \Sigma[R_i - E(R_i)] + \Sigma E(R_i).$$

7.7 Algebra of Stipulation (7.3)

Squaring both sides of the equal sign, and taking the expectation of each side, we have:

$$E(\Sigma R_i)^2 = E\{\Sigma[(R_i - E(R_i)] + \Sigma E(R_i)\}^2.$$

Now notice that the generic expression $(a + b)^2$ can be expanded to $a^2 + 2ab + b^2$. Thus, equating $\Sigma[(R_i - E(R_i)] = a$ and $\Sigma E(R_i) = b$, the right-hand side of the above equation can be expanded to:

$$E\{\Sigma[(R_i - E(R_i)]\}^2 + 2E\{\Sigma[(R_i - E(R_i)]\}\Sigma E(R_i) + E[\Sigma E(R_i)]^2.$$

However, the middle term of the above formulation vanishes because $E\Sigma[(R_i - E(R_i)] = \Sigma E(R_i) - \Sigma E(R_i) = 0$. [Note: $E\Sigma(R_i) = [\Sigma E(R_i)]$ and $E[\Sigma E(R_i)] = \Sigma E(R_i)$. Also, $E[\Sigma E(R_i)]^2 = [\Sigma E(R_i)]^2$. Appendix 7A shows some simple rules of operations on the expectations that are used in the above manipulations.] Thus, we find:

$$E(\Sigma R_i)^2 = E\{\Sigma[(R_i - E(R_i)]\}^2 + [\Sigma E(R_i)]^2. \tag{7.4}$$

The first term on the right-hand side can be expanded further to show:

$$E\{\Sigma[R_i - E(R_i)]\}^2 = \Sigma E[R_i - E(R_i)]^2 \\ + \Sigma\Sigma E\{[R_i - E(R_i)][R_j - E(R_j)]\} \tag{7.5a}$$

in which $j \neq i$, given $i = 0, 1, 2, \ldots, (t-1)$. [For the derivation of relationship (7.5a), see Appendix 7B(I).]

Notice further that the expression $E[R_i - E(R_i)]^2$ in the first term on the right-hand side of (7.5a) is the variance of R_i, whereas the expression $E\{[R_i - E(R_i)][R_j - E(R_j)]\}$ in the second term is the (auto)covariance between R_i and R_j, $i \neq j$. So, let us write these terms in more expressive ways: $E[R_i - E(R_i)]^2 \equiv \text{Var}(R_i)$; and $E\{[R_i - E(R_i)][R_j - E(R_j)]\} \equiv \text{Cov}(R_i, R_j)$. Thus, (7.5a) can be compactly written as:

$$E\{\Sigma[R_i - E(R_i)]\}^2 = \Sigma \text{Var}(R_i) + \Sigma\Sigma \text{Cov}(R_i, R_j), \\ j \neq i, \text{ given } i = 0, 1, 2, \ldots (t-1). \tag{7.5b}$$

We see that the first term on the right-hand side is the sum of variances for all i and the second term, the sum of all autocovariances (for all combinations of i and j, $i \neq j$). [The variance–covariance matrix in Appendix 7B(II) gives a more readily comprehensible picture of relationship (7.5b).]

Substituting the right-hand side of (7.5b) for the corresponding term in (7.4), and further substituting the results in (7.3), we have:

160 · **Regulation of Populations**

$$E(\Sigma R_i)^2 = \left[\Sigma \text{Var}(R_i) + \Sigma\Sigma \text{Cov}(R_i, R_j)\right] + \left[\Sigma E(R_i)\right]^2 < \infty. \quad (7.6)$$

We see that stipulation (7.6, hence 7.3) is made up of two independent components: $[\Sigma E(R_i)]^2$ and $[\Sigma \text{Var}(R_i) + \Sigma\Sigma \text{Cov}(R_i, R_j)]$ for $i = 0, 1, 2, \ldots, t-1$, $(j \neq i)$, each of which must be finite no matter how large t becomes. Let us write them down explicitly:

$$[\Sigma E(R_i)]^2 < \infty, \quad (7.6\text{-i})$$

$$\Sigma \text{Var}(R_i) + \Sigma\Sigma \text{Cov}(R_i, R_j) < \infty. \quad (7.6\text{-ii})$$

We may call (7.6-i) and (7.6-ii), respectively, the first and second requirements (stipulations) for the population to be in a persisting state. But what does each requirement signify ecologically?

7.7.1 Ecological Significance of the First Requirement (7.6-i)

Expression (7.6-i) can be written in a little more digestible form. Let $\mu_R(t)$ be the average of the series $\{E(R_i)\}$ over the period of length t such that $\mu_R(t) = [\Sigma E(R_i)]/t$, or $[\Sigma E(R_i)] = t\mu_R(t)$. Thus, the first requirement implies that $[t\mu_R(t)]^2 < \infty$, i.e. finite-valued, no matter how large t becomes. This in turn means that $\mu_R(t)$ must tend to zero to compensate for t increasing indefinitely: or more precisely, $\mu_R(t)$ must be infinitesimal of the same order as $t^{-1} (\equiv 1/t)$. Conversely, what if the mean $\mu_R(t)$ did not converge to zero as t increased? Recalling relationship (7.2a), the resultant series $\{X_t = \Sigma R_i + X_0; i = 0, 1, 2, \ldots, t-1\}$ would exhibit an increasing (or decreasing) trend without bound. In other words, the first requirement is to prevent the population from having an unbounded (unrestricted) trend (increasing or declining) so as to avoid reaching a dangerous level of extinction in a broad sense. [Note: Nothing is unusual about a declining population becoming extinct. But what about one increasing without bound? This happens only mathematically. Ecologically interpreted, the population concerned may destroy its habitat, leading eventually to its suicidal crash, a possibility with the present-day human population, long after the first warning by the Reverend Thomas Robert Malthus.]

Now, it is important to recognize that to meet the first requirement is necessary but not sufficient to conform to stipulation (7.3). Thus, the consideration of the second requirement is in order: in fact, this is the very essence of population regulation.

7.7.2 Second Requirement (7.6-ii) as the Essence of Population Regulation

Now that a variance is always positive by its definition, the sum $\Sigma \text{Var}(R_i)$ increases without bound as the length t of the series $\{R_t\}$ increases. So, to meet requirement (7.6-ii), the sum of the autocovariances $\Sigma\Sigma \text{Cov}(R_i, R_j)$ must be so negative as to nullify the positive and unbounded sum $\Sigma \text{Var}(R_i)$. In other words, the series $\{R_i\}$ must accordingly be autocorrelated. [Note: Inasmuch as the covariance $\text{Cov}(R_i, R_j)$ is called the 'autocovariance', the correlation between R_i and R_j, say $\text{Cor}(R_i, R_j)$, is called the 'autocorrelation'. These are related by the usual (Pearson) formula: $\text{Cor}(R_i, R_j) = \text{Cov}(R_i, R_j)/\{\text{Var}(R_i)\text{Var}(R_j)\}^{1/2}$.]

But what does a negative autocorrelation signify ecologically? It means that a high (positive) R_i, indicating an increase in population, is countered by a low (negative) R_j, indicating a population decrease, or vice versa. In other words, requirement (7.6-ii) is a precise way to define the notion of population regulation, commonly conceived as: 'When up, it must come down, and when down, it must go up'. Then, it is obvious that the 'regulation' is a necessary requirement for the persistent state of population but is not sufficient on its own: it must be supplemented by requirement (7.6-i).

With the above definition of population regulation in mind, we are ready to get into the details of the nature of population regulation. However, rather than going straight to it, I suggest that we look at a population in the mode of random walk as a typical example of unregulated processes.

7.8 Random Walk as Unregulated Processes

Recall the two aspects of the ensemble of simulated random-walk paths in Figure 7.1c: (a) apparent symmetry of the ensemble about the level on which $X_t = 0$; (b) apparently unbounded deviations of the ensemble about the level 0.

Aspect (a) as a sign of compliance with stipulation (7.6-i). Recall that, in the simulation, R_i was assumed to be an iid random number with mean $E(R_i) = 0$ for all i. This means that the expression $[\Sigma E(R_i)]^2$ simply vanishes to comply with stipulation (7.6-i). Thus, the apparent symmetry in the ensemble of random-walk series in Figure 7.1c is a manifestation of the compliance. More precisely, the ensemble mean of these series at a given t (i.e. the vertical sum of the R_t's at the given t, divided by the number of series in the ensemble, or the ensemble size) must converge to 0 as the ensemble size increases.

Incidentally, you might have wondered what would happen to a random-walk path if the mean $E(R_i) \neq 0$ were to violate (7.6-i). We would see that, for a positive or negative $E(R_i)$, the series concerned would wander around with a tendency to drift upwards or downwards. Accordingly, the ensemble of random-walk paths (Figure 7.1b or c) would be tilted up or down with the origin of the ensemble as a pivotal point.

Aspect (b) as a sign of non-compliance with stipulation (7.6-ii). Recall that, in a random-walk series, R_i is uncorrelated with R_j for $i \neq j$, i.e. the covariances $Cov(R_i, R_j)$ are identically zero for all $i \neq j$. That is, the second term on the (7.6-ii) (i.e. the sum of all covariances) vanishes, leaving the sum $\Sigma Var(R_i)$ alone, and the sum increases without bound as t increases. This attribute is manifest in the unbounded deviation (from the level 0) of the ensemble of the random-walk series in Figure 7.1c.

Altogether, a random walk is a typical example of unregulated population processes that does not comply with the second requirement (7.6-ii) for persistence, even though the first requirement (7.6-i) may be met. In fact, any process that does not meet the second requirement would exhibit an unregulated path, leading the population to its eventual extinction in the broad sense: as already explained, a population increasing unboundedly would eventually crash to extinction.

I now move on to consider the processes that meet the requirement (7.6-ii), i.e. the processes that are supposed to be regulated. In general, two classes of these processes are conceivable: one is regulated density-dependently, and the other density-independently. Let us look at the former class first.

7.9 Density-Dependent Regulation

Figure 7.2a shows an example of stochastic population series $\{X_t\}$ that model (6.1a) in Chapter 6 generates under the influence of the exogenous factor ε_i on parameter R_m, i.e. $(R_m + \varepsilon_i)$, $i = 0, 1, 2, \ldots (t-1)$. So, this stochastic version of model (6.1a) is written explicitly as:

$$R_t = (R_m + \varepsilon_t) - (h+1)\{\ln[1 + (c/h)\exp(X_t)]\}. \quad (7.7a)$$

Further, recalling that $X_t = R_{t-1} + X_{t-1}$ as defined, model (7.7a) can be expressed in the form ready for computing X_t:

$$X_t = X_{t-1} + (R_m + \varepsilon_{t-1}) - (h+1)\{\ln[1 + (c/h)\exp(X_{t-1})]\}. \quad (7.7b)$$

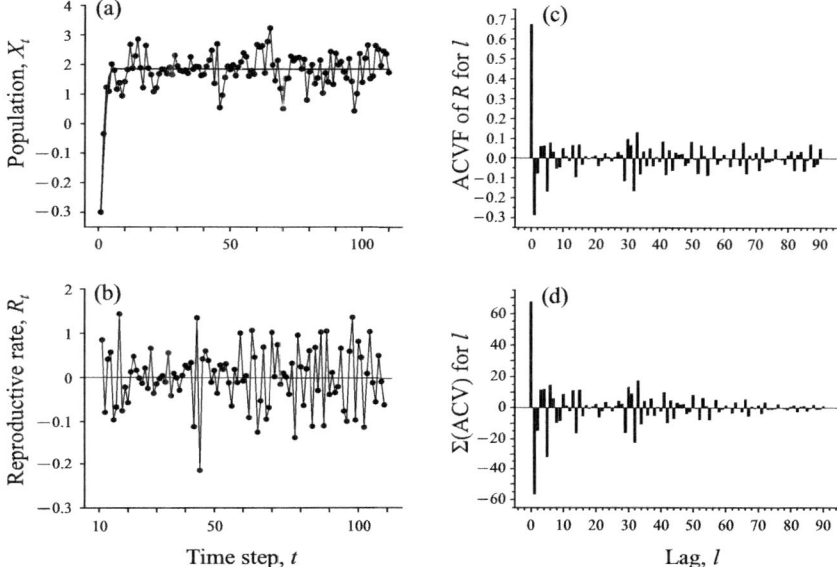

Figure 7.2 Graphical analysis of a population series to comply with stipulations (7.6).

In actually generating the Figure 7.2a series, I conveniently chose the set of (endogenous) parameter values: $R_m = 2$, $h \to 0$, and $c/h = 1$. As for the random series $\{\varepsilon_{t-1}\}$, I used the one in the first row of Figure 7.1a, which is an iid series, normally distributed with 0 mean and 0.25 variance.

Now, in fact, the above set of (endogenous) parameter values has generated the density-dependent reproduction curve in Figure 6.2, marked ($c = h \to 0$). In other words, this reproduction curve is the basis for generating the stochastic series $\{X_t\}$ in Figure 7.2a. In this sense, the $\{X_t\}$ is a stochastic density-dependent process. Let us analyse its behaviour.

In the same graph, I drew a solid curve (line, rather) generated by (7.7b) without the random influences $\{\varepsilon_{t-1}\}$. We see that, after about 5–6 time steps, the curve reaches the equilibrium level $X^* = 1.8546$ and stays there (horizontally) for good. [Note: Formula (6.3a) in Chapter 6 has shown how to calculate X^*, given a set of values for $(R_m, h, c/h)$.] Correspondingly, the superimposed stochastic series $\{X_t\}$ fluctuates in a haphazard manner but quite tightly about the equilibrium level: that is, the population appears to be regulated about (the endogenous level) X^*. In other words, after its initial (increasing) section, the stochastic series seems to have settled in a persistent state, meaning that it must have met stipulation (7.6). But how can we confirm that it has? This could be done by actually estimating (evaluating) the two terms on the right-hand side of (7.6) out of the generated series $\{X_t\}$.

Now that both terms are expressed in terms of R_i $(= X_{i+1} - X_i)$, series $\{X_t\}$ in Figure 7.2a is converted to $\{R_t\}$ in Figure 7.2b, excluding the first 10 time steps to ensure that $\{X_t\}$ has reached the equilibrium level. Thus, using series $\{R_t\}$, let us evaluate (estimate) the sum of R_i, in the interval from $i = 10$ to 110, as well as the sum of the variances and autocovariances to see if they comply with stipulation (7.6). However, as Murphy's Law predicts, things do not go as smoothly as we would wish: we encounter a conceptual problem that needs to be resolved.

The problem is that stipulation (7.6) for persistence is defined in terms of the expectations, i.e. the means $E(R_i)$ and the autocovariances $Cov(R_i, R_j)$, including the variance in which $i = j$. But, as already explained, these expectations are (conceptually) supposed to be the averages of the ensemble of (infinitely many) series with common statistical characteristics. However, we have only one series of observations in Figure 7.2b. How can we estimate these expectations? It is sort of like attempting to estimate the expected means and covariances of the weights of growing babies using just one series of observations with a single baby. Fortunately, in the present theoretical study, we can get around this problem if the process concerned is considered to be 'stationary', as explained below.

7.9.1 Stationary Processes

If certain moments of the series of random variables (especially its mean and variance) are time-invariant, the series is said to be stationary in these moments. The stochastic (solid–circle) series $\{X_t\}$ in Figure 7.2a, after it has reached the equilibrium level, is an example: also, the converted series $\{R_t\}$ in Figure 7.2b is stationary. You can actually confirm this if you extend (by more calculations) the series and pick a number of arbitrary sections of an adequate length to calculate the sample mean and variance of each section. You would find these moments to be more or less the same among the sections, or more precisely their differences to be statistically insignificant. In other words, an ensemble of these sections is comparable to a collection (population) of babies after they have reached adulthood, at which point their weights would no longer change (although this is not quite true in reality). It follows that sample moments

7.9 Density-Dependent Regulation · 165

of a single stationary series (of adequate length) are considered to be estimates of the corresponding expectations (as ensemble averages).

A pragmatic importance of the notion of stationarity in population ecology is that it enables (allows) us to estimate the mean and covariances from a single series of observations with an adequate length. [Note: These moments are often referred to as 'time averages', to be distinguished from 'ensemble averages'.] There is a caveat, however: an observed series may have a short-term trend such as a cyclic population series, like the wildlife's 10-year cycle. Then, in using sample mean or autocovariances (as time averages) to estimate the respective expectations (as ensemble averages), the length of the observed series must be sufficiently long to accommodate an adequate number of these cycles. [Note: This is a real problem with a species like spruce budworm with an average cycle length of about 30 years. When studying this species, my colleagues and I were lucky enough to have inherited a set of data for one cycle from our predecessors in addition to our own. Notwithstanding, we could not estimate these moments in a meaningful way.]

In reality, it is not easy to determine if the observed series can be treated as stationary. This is a big issue on its own and is beyond my expertise. For a rigorous treatment of the issue of stationarity, consult a textbook of stochastic processes; in particular, look into the concept of 'ergodicity'. Mind you, a statistical analysis is not the only way to study population ecology. Nonetheless, familiarizing yourselves with certain theoretical knowledge would help you deal with actual data from field or experimental studies.

With the foregoing perception of stationarity, we are now ready to test the persistent state of the simulated series $\{R_t\}$ in Figure 7.2b. The pragmatic implication of the method described below is that it should be applicable to observed series in actual field or laboratory studies, provided that the series are sufficiently long and appear to be stationary. [Note: It would be a good idea for you to generate a variety of stationary stochastic series, using theoretical models like (4.9) and (7.7b) to get acquainted with the patterns that comply with the stationarity concept.]

7.9.2 Testing the Persistent State of Observed Series

The R_t-series in Figure 7.2b fluctuates (albeit haphazardly) about the level (solid line) at the average of -0.0009 or practically 0. This means that, having reached the equilibrium level, series $\{X_t\}$ under the exogenous influences in Figure 7.2a exhibits practically no trend. In other words, the generated population series meets the first requirement (7.6-i) for a persistent state. So, let us examine whether the R_t-series in Figure 7.2b complies with (7.6-ii), i.e. $\Sigma \text{Var}(R_i) + \Sigma\Sigma \text{Cov}(R_i, R_j), j \neq i$, stays finite no matter how long the observation is made.

For this purpose, let us first calculate the autocovariances $\text{Cov}(R_i, R_j)$ for the lag $l = |i-j|$. [As noted in Appendix 7B(II), the difference $|i-j| = l$ is called the 'lag' of an autocovariance.] The calculation procedure can be readily explained in terms of the symmetric $t \times t$ variance–covariance matrix given in Appendix 7B(II). In particular, we consider the case in which $t = 100$ to match the length of the R-series in Figure 7.2b. [To follow the calculation procedures easily, I suggest that you use this matrix as a road map.]

To begin, look at the main diagonal (the diagonal series of the elements running across the matrix from the upper left corner down to the bottom right corner). Because it is made up of the elements for the lag $l = 0$, every element of the diagonal is a variance. Likewise, all the elements of the pair of the first off-diagonal series, i.e. immediately adjacent to the main diagonal (one above and the other below) are the autocovariances for lag $l = 1$. In general, the elements of the pair of l-th off-diagonal series are covariances for lag l.

Now that the R_t-series in Figure 7.2b is known to be stationary, translating the series into the matrix, all the elements in the l-th diagonal ($l = 0, 1, 2, \ldots, 100$) are (expected to be) of the same value. So, we can estimate this value by calculating the (sample) autocovariance for lag l, using the R_t-series that are realized in the simulation.

I calculated the covariances of the R_t-series for lags $l = 0, 1, 2, \ldots$, up to 90, and plotted them against l as the bar graph in Figure 7.2c. [Note: These bar graphs are collectively called the autocovariance function (ACVF) in that it is a function of l. How do you calculate the ACVF? See Appendix 7C.] With this ACVF chart in Figure 7.2c, we are ready to evaluate the terms on the left-hand side of inequality (7.6-ii) as functions of l.

Notice that, in the matrix in Appendix 7B(II), the number of elements in each of the diagonals is given by $(100 - l)$: for example, in the main diagonal, the number is 100 because $l = 0$. However, because the matrix is symmetric, all of the other diagonals come in pairs: one above and the other below the main diagonal. Therefore, the number of elements in each pair of diagonals for a given l is equal to $2 \times (100 - l)$. Thus, there are 2×99 elements in the pair of diagonals for $l = 1$; 2×98 for $l = 2$; and so on to 2×10 for $l = 90$. Then, I multiplied the bar graph for l in

Figure 7.2c by the corresponding number of elements, and the results (acronym: ΣACV for l on the vertical axis) are plotted against l in Figure 7.2d. [Note: The bar graphs converge to 0 as l increases and hence I did not bother to show those for the lags beyond 90.]

We see that the first bar in Figure 7.2d estimates the sum $\Sigma \text{Var}(R_i)$ as the first term on the right-hand side of (7.6-ii), which turned out to be $100 \times 0.6744 = 67.44$, because $\text{Var}(R_i)$ is estimated to be 0.6744 (as the first bar in Figure 7.2c). The sum of all other bars estimates the sum $\Sigma\Sigma\text{Cov}(R_i, R_j)$ as the second term on the right-hand side of (7.6-ii), which turned out to be -66.34. Thus, the total sum (of the elements in the matrix) turns out to be $67.44 - 66.34 = 1.1$, which easily meets the second requirement (7.6-ii) for persistence.

[Note: I strongly recommend that you try some simulations, using stochastic model (7.7) with varying parameter values and the statistical characteristics of the exogenous influences, to see whether they comply with stipulations (7.6).]

Now, let us delve into the precise nature of density-dependent regulation.

7.10 Precise Nature of Density-Dependent Regulation

7.10.1 Fundamental Nature

As is clear in the foregoing analyses, the fundamental nature of density-dependent regulation is to maintain the population about its equilibrium density X^* as depicted in the density-dependent reproduction curves in Figures 6.1 and 6.2. However, as already implicated, the density dependence of these curves has no control over the effect of a fluctuating environment. In other words, density-dependent regulation, under an adverse environmental effect, cannot maintain X^* within an appropriate range to comply with stipulation (7.6-i). Let us see what could happen then.

7.10.2 General Cases of Density-Dependently Regulated Populations

To simplify my argument without compromising generality, I use model (4.9) in Chapter 4 as it is algebraically easier to handle than (7.7). To repeat it for convenience:

$$R_t = (R_m + \varepsilon_t) - \exp[(C + \zeta_t) + X_t]. \qquad (4.9 \text{ rpt})$$

168 · Regulation of Populations

Here I consider cases in which the series of exogenous influences $\{\varepsilon_t\}$ and $\{\zeta_t\}$ are no longer iid random series but exhibit trends. Furthermore, I consider two (sub)cases: one in which $\{\zeta_t\}$ is active while $\{\varepsilon_t\}$ is asleep, and vice versa in the other.

Case 1. Consider that some environmental conditions reduce the supply of food. However, a reduction in food supply may imply that the area (s), which contains minimum sufficient food, would increase. This in turn implies an increase in parameter $c = s(1 - k)$, i.e. an increase in $(C + \zeta_t)$ in which $C = \ln(c)$. But, as already discussed in Chapter 4, an increase in C causes a lateral shift of the reproduction curve to the left, meaning a decrease in X^*. Nonetheless, the lateral shift of the reproduction curve would not change its slope at X^*, which means no change in the dynamical pattern of the population about X^*. In other words, the population can still be density-dependently regulated about the decreasing X^* in the same manner regardless of the lateral position of X^*. In an extreme situation, the population would be regulated about the decreasing X^* even to the final moment of extinction.

Case 2. A decrease in food supply may also cause a reduction in the (potential) reproductive rate $(R_m + \varepsilon_t)$, which in turn causes a decrease in the slope (steepness) of the reproduction curve at X^*. This further means that density-dependent regulation would become less tight. In an extreme situation in which the sum $(R_m + \varepsilon_t)$ becomes close to 0, the population may behave increasingly more like a random walk. However, the density-dependent reproduction curve would prevent the loosely regulated population from drifting upwards in the region above X^*, while allowing a downward drift below X^*. Under these circumstances, the likelihood of an accidental extinction would increase. This situation may be considered to be a depiction of the precarious status of some endangered species: in short, their fate would be predictably unpredictable.

To sum up, density-dependent processes that comply with requirement (7.6-ii) can ensure the population series $\{X_t\}$ to be regulated about the equilibrium density X^*. However, the density dependence is defenceless against the adverse effect of environmental changes on X^*. In other words, density-dependent regulation, i.e. compliance to stipulation (7.6-ii), is necessary but not sufficient to maintain the population in a state of persistence under changing environmental conditions: it must also comply with the first requirement (7.6-i), i.e. the elimination of a long-term trend.

So much for the nature of density-dependent regulation as the basis for the biotic-factor school, but how about density-independent regulation, the idea that the climatic-control school in effect advocated during the Great Debate?

7.11 Density-Independent Regulation

To talk about this issue, we need considerable imagination, or more like fantasies beyond the level of ordinary scientific inquiries. [Digression: A long ago, I was watching a science programme on TV in which the American cosmologist Professor Carl Sagan was the guest. The host interviewer asked Sagan what he thought about the book *Chariots of the Gods* by Erich von Däniken, in which the author quite seriously attributed some mysterious structures and objects in various parts of the world to the activities of extraterrestrial aliens. After a little pause, straight-faced Sagan replied: "Lack of imagination, I guess."]

Admitting that so aptly put by Sagan, I nonetheless went to see Captain James Kirk of the US Starship Enterprise at its Hollywood base. I asked him a favour to do two things for me: first, to go out into outer space where no man has gone before to find an alien species whose reproductive activities are governed solely by certain climatic conditions; and second, using Hollywood high tech, to build a rearing chamber in the starship, programmed to control the climate inside according to stipulations (7.6) such that the population of the alien species could persist without a density-dependent reproductive process.

The program I had in mind was: $R_i = \varepsilon_i - \varepsilon_{i-1}$ in which the $\{\varepsilon_i\}$ was an iid series with mean $E(\varepsilon_i) = 0$ and variance $Var(\varepsilon_i) = \sigma_\varepsilon^2$ for all i. Then, by relationship (7.2a) for $i = 1, 2, 3, \ldots, (t-1)$ and setting $X_0 = 0$ conveniently, we find:

$$X_t = \Sigma R_i = \Sigma(\varepsilon_i - \varepsilon_{i-1})$$
$$= (\varepsilon_1 - \varepsilon_0) + (\varepsilon_2 - \varepsilon_1) + (\varepsilon_3 - \varepsilon_2) + \ldots + (\varepsilon_{t-2} - \varepsilon_{t-3}) + (\varepsilon_{t-1} - \varepsilon_{t-2})$$
$$= \varepsilon_{t-1} - \varepsilon_0.$$

In other words, the alien species' population $\{X_t\}$ should be equal to the series of the iid random numbers $\{\varepsilon_{t-1} - \varepsilon_0\}$. However, as already pointed out and illustrated in Figure 7.1a, a series of iid random numbers is bounded such that the above population series $\{X_t\}$ should meet both stipulations (7.6-i) and (7.6-ii), i.e. the population should be in a persistent state. Captain Kirk became interested in my proposed project, built the rearing chamber equipped with the programmed control system, and set off. After a long while, I received a sorry message from him via the star-mail saying: 'Found the species. But the experiment failed!' The main cause of failure was the following.

No matter how well it was designed and built, the control system was unavoidably subject to small random (uncorrelated) errors, say $\{\eta_i\}$, the lower-case Greek 'eta'. As a result, the preprogrammed $R_i = \varepsilon_i - \varepsilon_{i-1}$ was in error as much as η_i such that what actually happened was:

$$X_t = \Sigma R_i = \Sigma[(\varepsilon_i - \varepsilon_{i-1}) + \eta_i] = \varepsilon_{t-1} - \varepsilon_0 + \Sigma \eta_i, \quad i = 1, 2, \ldots, t-1, \tag{7.8}$$

in which the sum $\Sigma \eta_i$ (as in the Figure 7.1 examples) is an uncontrolled random walk that failed the expensive project.

However, you might wonder what the series $\{X_t\}$ actually looks like. Its main path is the random walk set by the sum $\Sigma \eta_i$, but X_t deviates from it as much as $(\varepsilon_{i-1} + \varepsilon_0)$. That is, series $\{X_t\}$ fluctuates at random (nonetheless mostly within a certain bounded range) about the uncontrollably wandering main path. [You can find this yourself by simulation. If too lazy to do it, look at figure 1.8 of my book (Royama, 1992, p. 29).] In other words, the population meets requirement (7.6-ii), i.e. it is regulated about the unregulated random walk path. In short, my hopeful program only created a fragile state of persistence that did not withstand the effect of random errors $\{\eta_i\}$.

The lesson here is as follows. In the real world of ecology, density-independent regulation is too fragile to ensure a population to be in a state of persistence against ever-fluctuating environmental influences. Does this mean that the contention of the climatic control school should plainly be rejected? Not quite yet, as I now argue.

7.12 Logical Problem in Climatic-Control Theories

As I mentioned at the beginning of the present chapter, the very basis of the contention by the climatic control school is the claim that the variation in an insect population is often highly correlated with the variation in a climatic factor. Although I have not seen an actual case that is statistically substantiated by observation during my research career, I know it is a widespread belief among entomologists, particularly among those who are engaged in pest-control programmes.

The problem I see in the belief is not so much of a lack of substantiation by observations but a logical problem in interpreting the correlation. To put it more precisely, I have no problem with the assertion that

7.12 Logical Problem in Climatic-Control Theories · 171

climate influences the reproductive process of a biological population. I am sure it does: it need not be demonstrated inasmuch as we know for certain that a tropical species would not survive arctic weather and vice versa. Rather, the problem I see is a huge logical jump in deducing the belief that the correlation implies the climatic *regulation* of populations, as I reveal in a moment. However, as a preliminary step, let me show you that the correlation (between climatic variation and population fluctuation) could be real, as it is readily verifiable theoretically.

7.12.1 Reality of Correlation between Population Fluctuations and Climatic Influences

Recall the simulation in Figure 7.2a in which the population series $\{X_t\}$ was generated by model (7.7b) under the series of exogenous (density-independent) influences $\{\varepsilon_{t-1}\}$ taken from the first row of Figure 7.1a. It was also shown that series $\{X_t\}$ was density-dependently regulated because it was generated by one of the density-dependent reproduction curves in Figure 6.2 and because the sampled (stationary) section of the series met requirements (7.6). Now, using the stationary section after $t = 11$ in Figure 7.2a, let me correlate series $\{X_t\}$ with $\{\varepsilon_{t-1}\}$. Figure 7.3 shows the result. Wow, the correlation is nearly perfect! But how this can happen? It turns out to be very simple in principle.

As demonstrated in a few previous chapters, the dynamics of the (single-species) population series $\{X_t\}$ depends on the slope (tangent) of the reproduction curve at the equilibrium point X^*, i.e. the point where the curve crosses the $(R_t = 0)$-axis. So, consider a reproduction curve that crosses the axis at the angle $-45°$, i.e. its tangent line at X^* is a slanting line with a $-45°$ angle. This tangent line can be considered to be a linear approximation of the original curve in the vicinity of X^*. However, in the R–X coordinate plane, the equation of the $(-45°)$ line is given by $R_i = X^* - X_i$. Then, we can generate the series $\{X_t\}$ by the following linear (stochastic) model:

$$R_i = X^* - X_i + \varepsilon_i, \quad i = 0, 1, 2, \ldots, (t-1).$$

Now, by definition, $R_i = X_{i+1} - X_i$. So, for $i = (t-1)$, we find $X_t - X_{t-1} = X^* - X_{t-1} + \varepsilon_{t-1}$, and hence $X_t = X^* + \varepsilon_{t-1}$. In other words, in the above linear model, the variation in X_t in the vicinity of X^* is perfectly correlated with the random series $\{\varepsilon_{t-1}\}$. In the following, I show that the above attribute of the correlation between a population series and climatic influences reveals a logical flaw in the climatic control theory.

7.12.2 Logical Flaw in the Climatic Control Theory

Legend has it that in some parts of Europe, white storks (*Ciconia ciconia*) deliver babies to their respective homes to maintain the human populations there. Apparently, as I have heard, this has even been evidenced by a significantly high correlation between the stork populations and the numbers of babies delivered in the City of Amsterdam. Don't disregard this story as a folklore fantasy: it is in effect the way the climatic control theory has interpreted the weather–population correlation. In the real world of bugs, though, it is not the norm that the baby bugs are transported to their respective patch of woods by wind and are dumped there to form a new generation each time. Instead, what we usually see is that some climatic factors influence the net (intergeneration) reproductive rate of the population already existing in the woods, and this rate in turn determines the size of the new generation. Nevertheless, the climatic control theory assumes, in effect and quite subconsciously, that climate *directly determines* the size of the population.

The reality is that, as shown in the foregoing section, a high correlation occurs when the slope of the reproduction curve at X^* is about $-45°$, which means that the *endogenous* form of the reproduction curve is nearly perfectly density-dependent regulated: 'perfectly' in that if X_i happened to be displaced from X^*, the density X_{i+1} in the following generation would return right back (or at least very near) to X^*. The slope of the reproduction curve used in the simulation (Figure 7.2a) is about $-41°$ at X^*, sufficiently close to $-45°$ for the realized correlation to be as high as 98.7% in Figure 7.3. Needless to say, however, as the slope deviates from $-45°$ in either way, the correlation deteriorates.

Altogether, we see that the effect of the climatic influences would visibly show up in an observed population series *only when the population is tightly (or near-perfectly) regulated about its equilibrium point by a density-dependent process*, the point that the climatic control school completely failed to recognize. I find it rather ironic that the very basis of their contention ends up supporting the view held by the opposing, biotic control school of thought.

However, this does not mean that the logic of the opposing school is adequate; I must point out that it is just as weak. This is because the biotic control school, like the other, tried to demonstrate its perception by observations, mainly by correlations, failing to strongly present their contentions with logical arguments. The foregoing investigations reveal the following weaknesses in particular.

7.12 Logical Problem in Climatic-Control Theories · 173

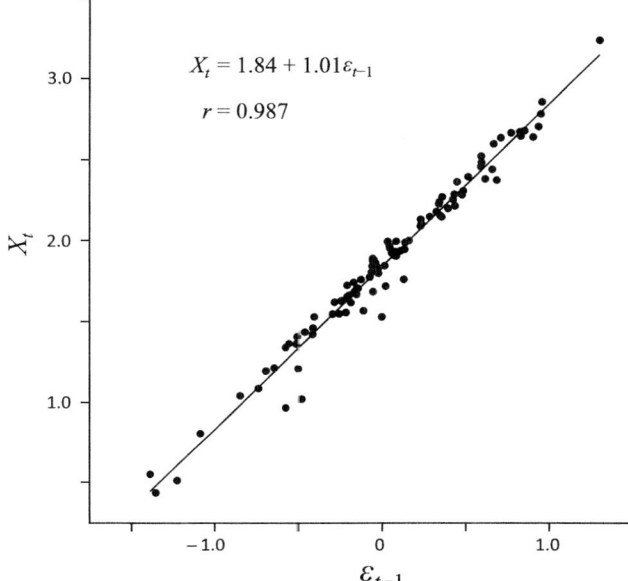

Figure 7.3 Correlation between the population series $\{X_t\}$ in Figure 7.2a and the series of random exogenous influences $\{\varepsilon_{t-1}\}$.

7.12.3 Logical Weaknesses in Biotic-Control School

As far as I have seen, the contention by the school during the Great Debates failed to bring out the following two issues that we have theoretically conceived in foregoing sections in the present chapter.

First, a population can be regulated either by a density-dependent process or by a density-independent process. However, the state of persistence in a density-independent process is fragile against a fluctuating environment. Hence, the likelihood of the occurrence of a density-independently regulated population in nature would be practically nil: it may be regulated density-independently but only temporarily. Second, although density-dependent regulation has a capacity to maintain the population about its equilibrium point X^* under a fluctuating environment, it has no capacity to maintain X^* within an appropriate range against an adverse effect of the environment.

After all, the Great Debates in the 1950s turned out to be like the Japanese saying: 'A big volcano rattles to get a little mouse out.' Worse still, it has created a myth which has led many ecologists to an unproductive attempt: testing density-dependent regulation by correlation, e.g.

by regressing the reproductive rate R_t on the population density X_t. I talk about this problem to conclude the present chapter.

7.13 Myth of Density-Dependent Regulation

An attempt to test density-dependent regulation by regression would likely fail because of two sorts of problems: one is pragmatic and the other logical.

7.13.1 Pragmatic Problem

Our observations in the field are usually limited in length. Even if you worked hard (on a univoltine insect) right from the beginning of your research career until retirement, you would have just one series of length not many more than 35 data points, barely enough to see one spruce budworm cycle. [Digression: I knew a professor of entomology in eastern Canada who, during his active carrier, never had an opportunity to actually witness how devastating a budworm outbreak could be, even though outbreaks have been well documented to occur quite regularly every 30 years or so.] What is more, if the series under observation (over a limited period in time) was not adequately stationary, you would not find much by the regression analysis.

7.13.2 Logical Problem

Consider that you were lucky enough to be accessible to a long series of observations in your research laboratory, and were able to do the analysis of an R_t-on-X_t regression. You might hope that, if the coefficient of a (linear) regression was significantly negative, it would demonstrate that the reproduction curve was indeed density-dependent in the narrow sense, i.e. R_t as a decreasing function of X_t with a finite equilibrium density X^*. Right? Wrong! The reason is the following.

Consider that the observed series you have was somehow known to have been generated by a density-dependent reproduction curve, e.g. one of those reproduction curves in Figure 7.2. Now, these curves are functional relationships, and it is certainly true that any simulated (observed) series generated by one of them would be manifest in a negative regression.

The problem here is that a negative regression is a statistical relationship and in general does not imply the functional relationship of a

7.13 Myth of Density-Dependent Regulation

density-dependent reproduction curve in the narrow sense, i.e. a decreasing function of population density. For instance, random exogenous influences can be autocorrelated to render the series $\{R_t\}$ autocorrelated, and the autocorrelation can yield a negative regression, even if the rate R_t is functionally unrelated to (independent of) the population density X_t. The following example shows how this can occur.

Recall the failed starship experiment, an example of density-independent regulation by (7.8), i.e. $R_t = \varepsilon_t - \varepsilon_{t-1} + \eta_t$, which has already been shown to be autocorrelated to meet stipulation (7.6-ii). Also by (7.8), we know that $X_t = \varepsilon_{t-1} - \varepsilon_0 + \Sigma\eta_i$ for $i = 0, 1, 2, \ldots$, $(t-1)$. Recall further that the (linear, Pearson) regression coefficient is given by the quotient $\mathrm{Cov}(R_t, X_t)/[\mathrm{Var}(R_t)\mathrm{Var}(X_t)]^{1/2}$ in which the denominator is positive by definition. So, let's calculate the covariance in the numerator to find the following:

$$\begin{aligned}\mathrm{Cov}(R_t, X_t) &= \mathrm{Cov}[(\varepsilon_t - \varepsilon_{t-1} + \eta_t), (\varepsilon_{t-1} - \varepsilon_0 + \Sigma\eta_i)] \\ &= \mathrm{Cov}(\varepsilon_t, \varepsilon_{t-1}) - \mathrm{Cov}(\varepsilon_t, \varepsilon_0) + \mathrm{Cov}(\varepsilon_t, \Sigma\eta_i) \\ &\quad - \mathrm{Cov}(\varepsilon_{t-1}, \varepsilon_{t-1}) + \mathrm{Cov}(\varepsilon_{t-1}, \varepsilon_0) - \mathrm{Cov}(\varepsilon_{t-1}, \Sigma\eta_i) \\ &\quad + \mathrm{Cov}(\eta_t, \varepsilon_{t-1}) - \mathrm{Cov}(\eta_t, \varepsilon_0) + \mathrm{Cov}(\eta_t, \Sigma\eta_i).\end{aligned}$$

Because both ε_t and η_t are iid as assumed in the experiment, all terms, except $\mathrm{Cov}(\varepsilon_{t-1}, \varepsilon_{t-1})$, vanish; even $\mathrm{Cov}(\eta_t, \Sigma\eta_i) = 0$ because the sum $\Sigma\eta_i$ is over $i = 0-(t-1)$, which is void of η_t and hence uncorrelated with it. However, the term $\mathrm{Cov}(\varepsilon_{t-1}, \varepsilon_{t-1})$ is in fact the variance, $\mathrm{Var}(\varepsilon_{t-1}) = \sigma_\varepsilon^2$. Thus, we find:

$$\mathrm{Cov}(R_t, X_t) = -\sigma_\varepsilon^2.$$

We see that the regression coefficient, i.e. $-\sigma_\varepsilon^2/[\mathrm{Var}(R_t)\mathrm{Var}(X_t)]^{1/2}$, is negative because σ_ε^2 and $[\mathrm{Var}(R_t)\mathrm{Var}(X_t)]^{1/2}$ are both positive. But why isn't the regression zero, despite the fact that the rate of change in population is functionally independent of density? The reason is as follows.

Recall that the (starship) experimental series $\{X_t\}$ was density-independently regulated about its random walk path. The foregoing negative regression is a result of this attribute of regulation. However, the random walk path is a density-independent process, such that, as t increases, $\mathrm{Var}(X_t)$ increases without bound, and the regression coefficient, i.e. $-\sigma_\varepsilon^2/[\mathrm{Var}(R_t)\mathrm{Var}(X_t)]^{1/2}$, would eventually (in the limit $t \to \infty$) converge to zero. The problem here is that your observation is limited in length, and what you find in an observation with a limited length is nearly always a negative regression. In fact, the sample regression coefficient could be surprisingly close to -1 when the length of observation is as short as 20 time steps, although gradually decreasing as the length increases. However, even after as many as 100 time steps, it could still be noticeably (significantly) negative; see the examples of simulation I presented in figure 1.10 of my previous book (Royama, 1992).

Altogether, the negative regression in the starship example is a manifestation of the fact that the population was regulated about the random-walk path. But you cannot affirm by regression whether the (observed) population is density-dependently or density-independently regulated. After all, a correlation (hence a regression) relationship does not necessarily imply a functional relationship, let alone the underlying mechanism.

It is true that in a density-independent process the regression would eventually converge to zero, but you may not see it in your comparatively short research career: your great-great-grandchild may. In addition, a negative regression could occur even if density-independent regulation (along the random-walk path) is not as ideal as in the starship experiment: it could occur when the series $\{R_t\}$ happens to be so autocorrelated. That is, a negative regression would be as ubiquitous as (on-the-whole negative) autocorrelations among random exogenous influences could be.

Incidentally, besides the aforementioned correlations which are real, an observed correlation can often be completely coincidental, e.g. a significant correlation between a stork population and the number of babies delivered. This is called a 'spurious' 'or nonsense' correlation. The problem is: in some cases it may not be as obvious and may fool you.

The bottom line is that, practically, an attempt to verify a density-dependent regulation by regression would get you nowhere or mislead at best because, in general, a regression relationship does not imply a functional relationship, a point that has evidently been not well recognized among ecologists.

7.13.3 Utility of the Regression

Despite the foregoing recognition, one might still argue: that the starship experiment was proven to be fragile; that we would be unlikely to see it occur in nature; that, conversely, every population we observe must be density-dependently regulated; and that, therefore, the observed negative regression implies density-dependent regulation. However, this argument contains logically uninspiring redundancy: if one has already been sure that what was observed must be density-dependently regulated, what is the point of attempting to confirm it by the indecisive regression?

That being said, though, an R_t-on-X_t regression (using an observed population series) is a useful device for estimating (constructing) an endogenous reproduction curve as a function of population density. In other words, the regression is still an indispensable tool for *estimating*, not for *testing*, density dependence.

7.14 Concluding Remarks of Chapter

The foregoing investigations have all been based on analyses of single-species populations. However, in its natural environment, a species would rarely exist on its own: it preys on other species; it is preyed upon by others; and it competes with others. A species is almost always an element of a food web or a system of the webs. Under these circumstances, a simple regression of R_t on X_t would not work because the reproductive rate of a species that interacts with other species is not only determined by its own density, but also by the densities of the interacting species. This necessitates us to generalize reproduction curves into rather complicated reproduction surfaces, formed by plotting the reproductive rate against population densities of the (at least) two interacting species.

Notwithstanding, the principles established in the present chapter would serve as a trailblazer when we go through a more complex maze of reality, the multiple-species interaction systems, that I attempt to explore in the following two chapters.

Appendix 7A: Rules of Operations on the Expectations Used in the Present Chapter

1. Expectation of a constant, say c, is itself: $E(c) = c$.
    Corollary: Expectation of the expectation of a random number, say u, is itself because an expectation is constant. That is, $E[E(u)] = E(u)$.
2. Expectation of the sum of random numbers $u, v, w, \ldots$ is the sum of their expectations:

$$E(u + v + w + \ldots) = E(u) + E(v) + E(w) + \ldots$$

178 · Regulation of Populations

3. Expectation of the multiplication of *independent* random numbers u, v, w, ... is the multiplication of their expectations: $E(uvw...) = E(u) \times E(v) \times E(w) \times ...$
4. If the random numbers u, v, w, ... are not independent, how to treat them depends on the case. For instance, in the case of a pair of random numbers in a series, say u_i and u_j, that occur $i - j = l$ time steps apart, then, $E\{[u_i - E(u_i)][u_j - E(u_j)]\}$ is an autocovariance for lag l. In particular, if u_i and u_j happened to be independent, Rule 3 applies: $E\{[u_i - E(u_i)][u_j - E(u_j)]\} = E[u_i - E(u_i)] \times E[u_j - E(u_j)]$. Then, applying Rules 2 and 1: $E[u_i - E(u_i)] = E(u_i) - E[E(u_i)] = E(u_i) - E(u_i) = 0$. Likewise, $E[u_j - E(u_j)] = 0$. That is, the autocovariance vanishes.

Appendix 7B: Derivation of Relationship (7.5)

(I). Derivation of (7.5a)

Let the expression $[R_i - E(R_i)]$ be written compactly as u_i such that:

$$\{\Sigma[R_i - E(R_i)]\}^2 = (\Sigma u_i)^2$$
$$= (u_0 + u_1 + u_2 + \ldots + u_{t-1})$$
$$\times (u_0 + u_1 + u_2 + \ldots + u_{t-1}).$$

Now, the expression on the right-hand side of the above relationship can be expanded as:

$$u_0^2 + u_0(u_1 + u_2 + \ldots + u_{t-1})$$
$$+ u_1^2 + u_1(u_0 + u_2 + \ldots + u_{t-1})$$
$$\ldots\ldots\ldots\ldots\ldots\ldots\ldots\ldots\ldots\ldots\ldots\ldots$$
$$+ u_i^2 + u_i(u_0 + u_1 + \ldots + u_{i-1} + u_{i+1} + \ldots + u_{t-1})$$
$$\ldots\ldots\ldots\ldots\ldots\ldots\ldots\ldots\ldots\ldots\ldots\ldots$$
$$+ u_{t-1}^2 + u_{t-1}(u_0 + u_1 + u_2 + \ldots + u_{t-2}).$$

The sum of the above lows can be compactly written as:

$$\Sigma u_i^2 + \Sigma u_i(\Sigma u_j), \quad \text{given } i = 0, 1, 2, \ldots, t-1; j \neq i,$$

and writing the term $\Sigma u_i(\Sigma u_j) \equiv \Sigma\Sigma u_i u_j$, and rewriting u back to its original form, we have:

$$\{\Sigma[R_i - E(R_i)]\}^2 = \Sigma[R_i - E(R_i)]^2 + \Sigma\Sigma\{[R_i - E(R_i)][R_j - E(R_j)]\}$$

in which $j \neq i$, given $i = 0, 1, 2, \ldots, t - 1$. Taking the expectations of the terms on both sides, we have (7.5a).

(II). Variance–Covariance Matrix of the Right-Hand Side of (7.5b)

Now that $\{\Sigma[R_i - E(R_i)]\}^2 = \{\Sigma[R_i - E(R_i)]\} \times \{\Sigma[R_i - E(R_i)]\}$ for $i = 0, 1, 2, \ldots, t - 1$, the left-hand side of (7.5b) can be represented by the $t \times t$ square array of the elements, comprising all variances and covariances on the right-hand side of (7.5b). Hence, the array is called the 'variance–covariance matrix'.

The matrix contains all possible combinations of $Var(R_i)$ and $Cov(R_i, R_j)$ for i and $j \neq i$ each of which takes integer values in the range $[0, t - 1]$. [Note: the square brackets indicate that the two extreme values are included in the range. If excluded, then a parenthesis is used. For instance, if 0 was excluded, then $(0, t - 1]$.] For compactness sake, we may write the expressions $Var(R_i)$ as $\mathbf{V}(i)$, and $Cov(R_i, R_j)$ as $C(i, j)$. The integers i and j, respectively, designate row and column numbers, both of which are in the range $[0, t - 1]$ to form a $t \times t$ symmetric matrix. The set $\{\mathbf{V}(i); i = 0, 2, \ldots, t - 1\}$ comprises the main diagonal of the matrix, and all elements off the main diagonal comprise the set $\{C(i, j)\}$ which is arranged in the order of the difference (lag) $|i - j| = l$. In particular, the first, second, third, ..., and $(t-1)$-th diagonals below and above the main diagonal are those for $l = 1, 2, 3, \ldots,$ and $t - 1$. Because $|i - j| = l$, those covariance elements in the corresponding diagonals above and below the main diagonal are equal in value, i.e. $C(j, i) = C(i, j)$. Hence, the matrix is symmetric about the main diagonal. Clearly, then, there are t variances and $2(t - l)$ covariances for $l = 1, 2, 3, \ldots, t - 1$, as shown in the variance–covariance matrix below.

	$j = 0$	1	2	.	$t - 3$	$t - 2$	$t - 1$
$i = 0$	$\mathbf{V(0)}$	$C(0, 1)$	$C(0, 2)$	.	$C(0, t - 3)$	$C(0, t - 2)$	$C(0, t - 1)$
1	$C(1, 0)$	$\mathbf{V(1)}$	$C(1, 2)$	.	$C(1, t - 3)$	$C(1, t - 2)$	$C(1, t - 1)$
2	$C(2, 0)$	$C(2, 1)$	$\mathbf{V(2)}$	.	$C(2, t - 3)$	$C(2, t - 2)$	$C(2, t - 1)$
.	.	.	.	.	.	.	.
$t - 3$	$C(t - 3, 0)$	$C(t - 3, 1)$	$C(t - 3, 2)$	.	$\mathbf{V(t-3)}$	$C(t - 3, t - 2)$	$C(t - 3, t - 1)$
$t - 2$	$C(t - 2, 0)$	$C(t - 2, 1)$	$C(t - 2, 2)$	.	$C(t - 2, t - 3)$	$\mathbf{V(t - 2)}$	$C(t - 2, t - 1)$
$t - 1$	$C(t - 1, 0)$	$C(t - 1, 1)$	$C(t - 1, 2)$	.	$C(t - 1, t - 3)$	$C(t - 1, t - 2)$	$\mathbf{V(t - 1)}$

Appendix 7C: Calculation of an Autocovariance Function (ACVF)

It would be a tedious task to hand-calculate an ACVF. Luckily, we have PCs nowadays. Not so lucky is that some statistical software packages (e.g. Minitab) may not provide the ACVF package. Luckily again, though, many software programs offer an autocorrelation function (ACF) package. Because an autocorrelation coefficient is the Pearson correlation coefficient applied to the two data points of a series, say $\{u_i\}$, that are $l = |i-j|$ time steps apart, it takes the form:

$$\mathrm{Cov}(u_i, u_j) / [\mathrm{Var}(u_i) \mathrm{Var}(u_j)]^{1/2}.$$

Furthermore, for a stationary series $\{u_i\}$ in which $\mathrm{Var}(u_i) = \sigma_u^2$ for all i, the above expression simply becomes $\mathrm{Cov}(u_i, u_j)/\sigma_u^2$. However, because the calculation of a variance is a standard package in any software, you calculate the autocorrelation coefficients and multiply them with the square root of the variance term to get the ACVF.

8 · *Predator–Prey Interaction Processes*

8.1 Preamble

The foregoing investigations of single-species processes have been based on the assumption that the availability of food was invariant in time: if depleted by the end of each generation, it would have recovered by the beginning of the following generation. Here I consider that, in general, the prey population would be so depleted by predation that it may take a few generations to recover, a situation which in turn influences the dynamics of the predator population, and so on goes the predator–prey interaction process in time. Thus, a predator–prey process is a generalized single-species process in which the effect of food depletion and its recovery is explicitly accounted for. [Note: Although commonly called a predator–prey interaction, the predator category involves insect parasitoids which eat up their hosts eventually. In effect, a parasitoid acts as a (slow-killing) predator, and hence is called 'parasitoid' to be distinguished from a truly parasitic organism which would not consume its host.]

To begin the investigation into the interaction process, let us formulate a model of suitable form that incorporates the above attributes of the interaction processes into the net reproductive rate (intergeneration balance between births and deaths) of each species. I consider endogenous processes first and then subject them to exogenous influences.

8.2 Formulation of Endogenous Models of the Interaction Processes

8.2.1 Categories to be Considered

As discussed in Chapters 3 and 4, population processes in general can be classified into two categories: continuous-time and discrete-time processes. Here I consider the discrete-time category: each species involved has a distinct breeding season such that generations do not overlap each other. I do not consider continuous-time processes (in which breeding occurs at any point in time and hence generations are not distinct). The reason for this approach is as follows.

Most importantly, as has been exemplified in Chapter 4, a continuous-time process tends to hide details of the time-dependent aspects of the rate of change in population, whereas a discrete-time process can readily reveal them. Also, a continuous-time process is usually represented by a differential equation, as in the classical logistic models, i.e. models (3.1) in Chapter 3, or the well-known Lotka–Volterra model of predator–prey interaction. However, a differential equation, unless as simple as those mentioned above, tends to be cumbersome to handle analytically and computationally. This difficulty often necessitates simplification of the model at the expense of ecological reality: as a matter of fact, the Lotka–Volterra model is an example, as I show it in due course.

In contrast, a discrete-time process is represented by a difference (recurrence) equation which, like theoretical model (3.10), tends to be more readily structured to represent an ecological process. Moreover, as has been shown in Chapter 4, a discrete-time process tends to yield more varieties of dynamical (time-dependent) features of population processes and hence is more widely applicable to many species we actually observe in nature.

After all, a continuous-time process is a special case of the corresponding discrete-time process, as exemplified in Chapter 3. Thus, if we understood a given discrete-time process, we could infer what the corresponding continuous-time process would be like. Besides, the pragmatic advantage of a discrete-time model, built in the recurrence-equation form, is its ease of doing computer simulations.

8.2.2 Notations to be Used

Let X and Y stand, respectively, for the prey and predator (parasitoid) species whose densities at the t-th generation are denoted by x_t and y_t. Also, let r_t and r'_t be the (net intergeneration) reproductive rates of X and Y, respectively. Then, as before, we define: $r_t = x_{t+1}/x_t$ and $r'_t = y_{t+1}/y_t$. However, each of r_t and r'_t is a function of both x_t and y_t. To make this property of the reproductive rates explicit, we may write $r_t \equiv r_t(x_t, y_t)$ and $r'_t \equiv r'_t(y_t, x_t)$.

[Notes: X and Y are the names (or markers; not algebraic symbols) of the two species and are not italicized, to be distinguished from the italicized counterparts, X and Y, denoting their (log-transformed) densities, e.g. $X \equiv \ln(x)$ and $Y \equiv \ln(y)$. Note also that X and Y could be in the natural or common logarithm, and whenever necessary to distinguish between the two, I use 'ln' for the natural logarithm and '$\log_{10}$' (or just 'log') for the common logarithm. In particular, an ln-transformation is used in algebraic expressions (equations), whereas a log-transformation is used in graphical representations, unless otherwise specified. If either will do, it will not be specified.]

8.2.3 Formulation of the Models

As a generalization of single-species processes, it is natural to consider that model (4.1) of Chapter 4 would serve as a basic structure of the reproductive rate in both prey species X and predator species Y. [Note: Although model (4.1) is a special case of model (5.12) in Chapter 5, I use (4.1) mostly here because it is algebraically simpler to handle without losing generality in investigating the ecological principles of predator–prey interaction processes. I will talk about the utilities of the general model (5.12) later in this chapter.]

To adapt model (4.1) to the present two-species situation, I write it specifically for each species as:

$$r_t(x_t) = r_m \exp(-cx_t) \quad (4.1 \text{ for } X)$$

$$r'_t(y_t) = r'_m \exp(-c'y_t) \quad (4.1 \text{ for } Y)$$

where r_m (or r'_m) is the mean potential (maximum) reproductive rate of an individual in X (or Y) when there is no fellow competitors; and c (or c') is a measure of the intensity of intraspecific (within-population) competition in X (or Y). [Note: Every one of these parameters is positive and finite-valued. However, I consider that r_m and r'_m are both greater than 1 because, otherwise, the populations would go extinct, a trivial case in the present theme.]

Now notice that, in a predator–prey interaction process, model (4.1 for X) would stand as is when predator Y is absent, whereas (4.1 for Y) would stand as is when prey X is plentiful. But, in general, the reproductive rate of one species is affected by the abundance of the other species in a certain manner. Let the expression $p(y_t)$ designate this 'certain manner' of the effect of predator abundance (y_t) on prey X, and let $q(x_t)$ be the effect of prey availability (x_t) on Y. Then, it is natural to think that the reproductive rate $r_t(x_t, y_t)$ of X is proportional to $p(y_t)$, given x_t

(i.e. when x_t is fixed at a certain level) such that model (4.1 for X) is generalized to:

$$r_t(x_t, y_t) = [r_m \exp(-cx_t)] \, p(y_t). \tag{8.1a}$$

Likewise, the reproductive rate $r'_t(y_t, x_t)$ of Y in model (4.1 for Y) is proportional to $q(x_t)$, given y_t, and is written:

$$r'_t(y_t, x_t) = [r'_m \exp(-c'y_t)] q(x_t). \tag{8.1b}$$

Now, evidently, the reproductive rate of prey X, i.e. $r_t(x_t, y_t)$ in (8.1a) cannot exceed r_m because r_m is defined as the potential (maximum) reproductive rate of an individual prey when there is no predator. Thus, $p(y_t)$ should not exceed 1. Neither can the reproductive rate for the predator Y, i.e. $r'_t(x_t, y_t)$ in (8.1b) exceed r'_m because r'_m is defined as the maximum reproductive rate of Y, when X (Y's food) is plentiful. It follows that neither should $q(x_t)$ exceed 1. In the meantime, obviously, both p and q are positive functions. In other words, we recognize that $0 < p(y_t) \leq 1$ and $0 < q(x_t) \leq 1$. In the following, let us consider what the effects p and q would be like in concrete forms. To find them, it is necessary to take certain basic ecological attributes into account.

8.2.4 Ecological Attributes of the Effects $q(x_t)$ and $p(y_t)$

Let us look into $q(x_t)$ first. Quite clearly, the effect is proportional to the number of prey individuals that a predator can consume per unit time. This implies that the curve formed by $q(x_t)$ plotted against x_t should look like the profitability curve of Figure 1.1 in Chapter 1. It starts from the origin $x_t = 0$; increases monotonically but with an ever-decreasing rate of increase as x_t increases; and eventually converges asymptotically to the upper level = 1 because of the constraint $q \leq 1$. Then, the curve would obey the popular Disc Equation of C. S. Holling (1959), as referred to in relation to Figure 1.1, but with the asymptotic level = 1. However, for algebraic convenience, I use the alternative formula:

$$q(x_t) \equiv [1 - \exp(-b'x_t)]. \tag{8.2}$$

Appendix 8A shows the mechanism underlying formula (8.2). The ecological meaning of the parameter b' will be discussed in a moment.

Now, let us look into $p(y_t)$. Essentially, the $p(y_t)$ curve, the plot of $p(y_t)$ against y_t, should be a reversal of the $q(x_t)$ curve in shape, inasmuch as the reproductive rate (8.1a) of prey X is affected by the number eaten. In particular, the curve starts at 1 when there is no predator, i.e. $y_t = 0$, and decreases monotonically and asymptotically to 0 as y_t increases. Then, in the present context, the most appropriate formula to describe the above tendency would be:

$$p(y_t) \equiv \exp(-by_t), \qquad (8.3)$$

although parameter b generally differs from b'. Now, let us look into the ecological meaning of b' and b.

8.2.5 Ecological Meaning of Parameters b' and b

Parameter b' characterizes the effect of prey abundance $q(x_t)$ on the predator's reproductive rate, $r'_t(x_t, y_t)$ in (8.1b). In particular, the higher the b' value, the more readily does an individual predator collect prey per unit time (effort) and, hence, the more quickly does $q(x_t)$ approach 1 asymptotically. Thus, b' is a measure of the hunting efficiency (skill) of the predator.

Parameter b characterizes the effect of predation $p(y_t)$ on the reproductive rate of the prey, i.e. $r_t(x_t, y_t)$ in (8.1a). In particular, the higher the b value, the faster $p(y_t)$ decreases, and so does the prey's reproductive rate. Meanwhile, the higher the predator's consumption of food per unit time, the greater the impact on the prey's reproductive rate, quite naturally. Thus, b is a measure of the predator's per-capita consumption (requirement) of food.

8.2.6 Full Formulation of the Reproductive Rates

Substituting the expression $\exp(-by_t)$ in (8.3) for $p(y_t)$ in (8.1a), and the expression $[1 - \exp(-b'x_t)]$ in (8.2) for $q(x_t)$ in (8.1b), we have the full formulae to represent the reproductive rates of both species:

$$r_t(x_t, y_t) = [r_m \exp(-cx_t)] \exp(-by_t) \qquad \text{for prey X} \quad (8.1a \text{ full})$$

$$r'_t(y_t, x_t) = [r'_m \exp(-c'y_t)][1 - \exp(-b'x_t)] \qquad \text{for predator Y.}$$
$$(8.1b \text{ full})$$

[Note: These are identical in form with the predator–prey models that I developed in my earlier book (Royama, 1992; models (5.17), p. 194). The two versions differ only in notations.]

We see that there are six parameters altogether: (r_m, c, b) for X and (r'_m, c', b') for Y. Each of them has its own ecological meaning as defined above. However, from an algebraic point of view, we can reduce the number to four to simplify the analyses of the model's dynamical (time-dependent) aspects without loss of information on the ecological features of the interaction process.

8.2.7 Reducing the Number of Parameters

The trick is to temporarily write $cx_t \equiv u_t$, say, in (8.1a) such that $x_t = u_t/c$. Substituting u_t/c for x in (8.1b), we find: $b'x_t = b'(u_t/c)$. Likewise, letting $c'y_t \equiv v_t$ in (8.1b), we find $by_t = b(v_t/c')$ in (8.1a). Thus, algebraically, each of (b/c') and (b'/c) acts as a single parameter without affecting the model structure or affecting the dynamical features of the interaction process. In fact, the above manipulation can be interpreted as a simple linear transformation of the unit of measuring the population density of either species, e.g. counting the number of individuals in units of 10, 100, or in whatever is convenient for counting, e.g. insect eggs individually or en masse. Then, within the framework of the present theoretical investigation, we may conveniently set $c = c' = 1$ in (8.1) to 'standardize' the unit of measurement, so to speak. Thus, the system model (8.1a and b full) is simplified to:

$$r_t(x_t, y_t) = [r_m \exp(-x_t)] \exp(-by_t) \qquad \text{for prey X} \qquad (8.4a)$$

$$r'_t(y_t, x_t) = [r'_m \exp(-y_t)][1 - \exp(-b'x_t)] \qquad \text{for predator Y.} \qquad (8.4b)$$

We are now ready to look into the dynamics of system model (8.4) by simulation. Let us look into the endogenous features in the following before considering the effects of exogenous influences.

8.3 Simulation of the Dynamics of Predator–Prey Interactions

For this purpose, let me modify model (8.4) into ready-for-calculation formulae. Using the definitions $r_t \equiv x_{t+1}/x_t$ and $r'_t \equiv y_{t+1}/y_t$, we have:

$$x_{t+1} = x_t[r_m \exp(-(x_t))] \exp(-by_t) \qquad (8.5a)$$

$$y_{t+1} = y_t[r'_m \exp(-y_t)][1 - \exp(-b'x_t)]. \qquad (8.5b)$$

Thus, given certain arbitrarily chosen values for the four parameters, as well as the initial set of densities (x_0, y_0) at $t = 0$ (also arbitrarily chosen),

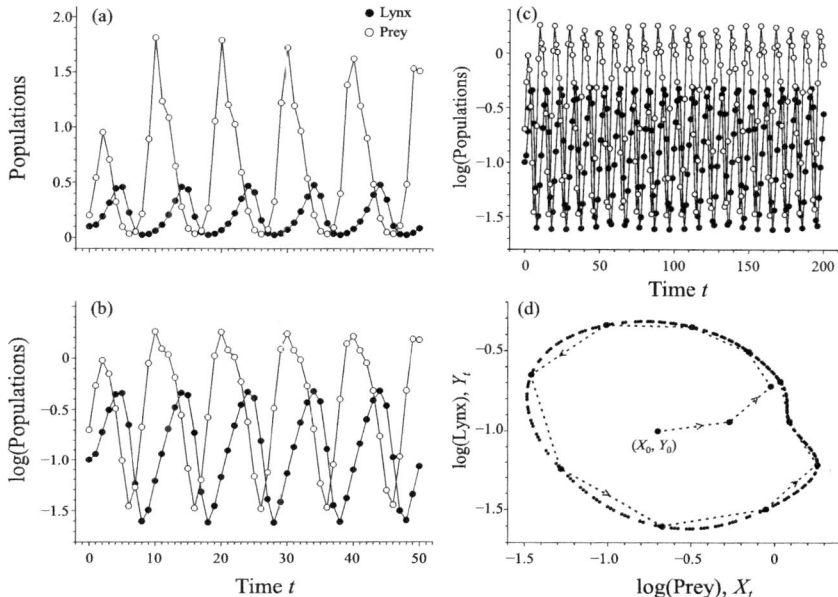

Figure 8.1 An example of simulated predator–prey interaction process, using models (8.5).

we can recursively calculate (x_1, y_1), (x_2, y_2), ..., and so on for as many generations (t) as you like.

Figure 8.1 shows an example of simulation, using formulae (8.5) with parameter values $(r'_m, b') = (2, 5)$ and $(r_m, b) = (6, 6)$. [Note: I chose these sets of parameter values to mimic the well-known Canada lynx (*Lynx canadensis*) 10-year cycle, as depicted by the pelt-harvest records from the Mackenzie River District of the Hudson's Bay Company: cf. figure 5.1b in my previous book (Royama, 1992, p. 170).]

Figure 8.1a is scaled in the usual (non-negative) counts of x_t (prey) and y_t (lynx). Both of them are $\log_{10}$-transformed in Figure 8.1b such that one unit of scale indicates a 10-fold change in population. Now, let us look at a few aspects of the simulated cycles in Figure 8.1b.

8.3.1 Phase Shift between Predator and Prey Cycles

We notice a phase shift between the prey and predator cycles: the predator cycle is lagged behind the corresponding prey cycle. This has been well-known ever since the advent of the classical Lotka–Volterra model, and as quoted in most ecology textbooks. [Note: This phenomenon is often referred to as a 'time delay', e.g. 'delayed density dependence'. Personally, I prefer 'shift' or 'lag'. This is because the expression 'delay' can be used more appropriately in another context, such as: food shortage in winter determines (influences) natality in spring.]

8.3.2 Asymmetry of a Cycle

Not so well-recognized is the asymmetry of a cycle. In the present example of Figure 8.1, it takes on average 6 units of t (6 years) for a predator cycle (solid circle) to increase from a trough to a peak, whereas it takes only 4 years to come down to the trough. In contrast, it takes 4 years for a prey cycle (open circles) to reach its peak and 6 years to come down. In other words, the predator cycle leans to the right, which in geometry is said to be 'skewed to the left' (or left-skewed), whereas the prey cycle is skewed to the right (leaning to the left). In fact, the left-skewed predator cycle is clearly exemplified in the observed lynx pelt-harvest records.

Unfortunately, we cannot confirm the right-skewed prey cycles in observations to be paired with the lynx cycle because no comparable data exist in the Hudson's Bay Company business statistics.

[Note: I have seen some articles in which a lynx pelt series is paired with a snowshoe hare harvest series to illustrate predator–prey cycles. Read those articles with caution, because the paired series as illustrated are, in all likelihood, not directly comparable in that they were statistics taken for different business purposes, or might have been from different geographical regions: see the remarks I made on p. 233 (middle paragraphs) in Royama (1992).]

Despite the above problem, the asymmetry of a cycle in the simulation makes good sense theoretically. It is the consequence of the comparatively large (as so assumed) difference between the two species in their potential reproductive rates ($r_m = 6$, $r'_m = 2$), as well as due to the comparatively high demand for food ($b = 6$) by the predator. In particular, starting with a low predator population (y_t), the prey population (x_t) increases quickly because of its high potential reproductive rate $r_m = 6$. As x_t increases, y_t follows, but only slowly because of its low potential rate $r'_m = 2$. As y_t increases to a sufficiently high level, it causes x_t to decrease, but it does so only slowly at first because y_t has not yet reached its peak level (full strength, so to speak). As y_t increases further, it eventually causes x_t to decrease in an accelerated rate. This in turn causes y_t to decrease as well, but it decreases quickly because its high demand for food ($b = 6$) makes it starve quickly in a diminishing food (prey) supply.

In short, the prey increases quickly and decreases slowly, whereas the opposite is true with the predator, and these tendencies result in skewness in their cycles in opposite directions.

8.3.3 Apparently Perpetual Oscillations

In the Figure 8.1a and b examples, the oscillations appear to continue without a sign of converging to an equilibrium point. In fact, these are so-called limit cycles, i.e. perpetually repeating the same pattern. But, how do we know that these cycles are limit cycles? Actually, as far as we see in Figure 8.1b, two adjacent cycles do not seem to be identical to each other. Nonetheless, if the series is extended further in Figure 8.1c, we do see certain regularities in the cycles; it even looks rather pretty aesthetically. It shows that, rather than the same cyclic pattern being repeated one after another, a number of consecutive cycles as a set is repeated regularly in a certain pattern. This tendency of limit cycles can be more readily recognizable in yet another type of graphical representation, called the phase-space representation in applied mathematics, to describe dynamical systems in physics.

8.3.4 Phase-Space Representation

In Figure 8.1d, I plotted Y_t, the $\log_{10}$(lynx population), against X_t, the $\log_{10}$(prey population), using the pair of series in Figure 8.1c. Starting at the point (X_0, Y_0), i.e. $t = 0$, at around the centre of graph (d), the track of the subsequent points (X_t, Y_t) for $t = 1, 2, 3, \ldots$, goes anticlockwise to converge to a closed elliptic orbit. [The orbit would be clockwise if you plot Y against X.] Also, if you started with point (X_0, Y_0) outside the ellipse (not shown here), you would see the track converge to the same orbit. In other words, the closed orbit is stable in that no matter where the series starts, it ends up converging to this closed orbit, which is the characteristic of limit cycles. Incidentally, what about the dimple on the right-hand side of the ellipse? This is due to the aforementioned asymmetry of the X and Y cycles that are more widely apart from each other at their peaks than at the troughs.

Now, let us look into some more varieties of dynamical pattern that the system model (8.4), or the formulae (8.5) for calculations, would generate.

8.4 Variation in Dynamical Patterns

As shown in Chapter 4, the single-species model (4.1) generates a variety of dynamical patterns about the equilibrium level, depending on the variation in the potential reproductive rate r_m: cf. Figure 4.6). This feature (dependence on r_m) resulted in the three major categories of patterns, namely: monotonic convergence to equilibrium as in Figure 4.6a; convergence with damped oscillations as in Figure 4.6b and c; and perpetual oscillations about an equilibrium level as in Figure 4.6d–h. In any case, the oscillations are invariably up and down alternating (i.e. saw-toothed) about the equilibrium level. As compared with these simple features, the variation in dynamical pattern in the predator–prey model (8.4), or the formulae (8.5) for calculation, depends on combinations in values of the four parameters (r_m, r'_m, b, and b').

The combinations generate many more varieties in dynamical pattern, and going through all possible combinations to see what might show up would be impractical. Nonetheless, the variation in model (8.4) can still be recognizable in three major categories more or less corresponding to those in Figure 4.6, namely: non-oscillating convergence to equilibrium; damped-oscillations converging to equilibrium; and perpetual, undamped oscillations. These are illustrated by examples in Figure 8.2 with the sets of parameter values (noted in each graph) that are conveniently chosen for illustration purposes. The graphs on the left are the log-transformed population series, and those on the right are their phase-space representations.

For simplicity, the (r_m, r'_m) values are fixed at (2.7, 2.0) in all these series. This is because the variation in the dynamical pattern of a predator–prey process (that differs from the variety in a single-species process) depends primarily on the values of the parameters (b, b'); b being a measure of the predator's requirement (demand) for food (prey) and b' its hunting efficiency (skill). In the following, I analyse the patterns exhibited by the Figure 8.2 examples in terms of the variations in their (b, b') values.

8.4.1 Analysis of Dynamical Patterns

In Figure 8.2a, I chose (b, b') = (6.0, 5.0) to indicate that the predator is a demanding and efficient hunter. This example exhibits limit cycles, similar to the simulated lynx cycles in Figure 8.1, although the Figure 8.2a$_1$ cycles are less skewed. This is because the difference in the potential reproductive rates between the two species is reduced to (r_m,

8.4 Variation in Dynamical Patterns · 191

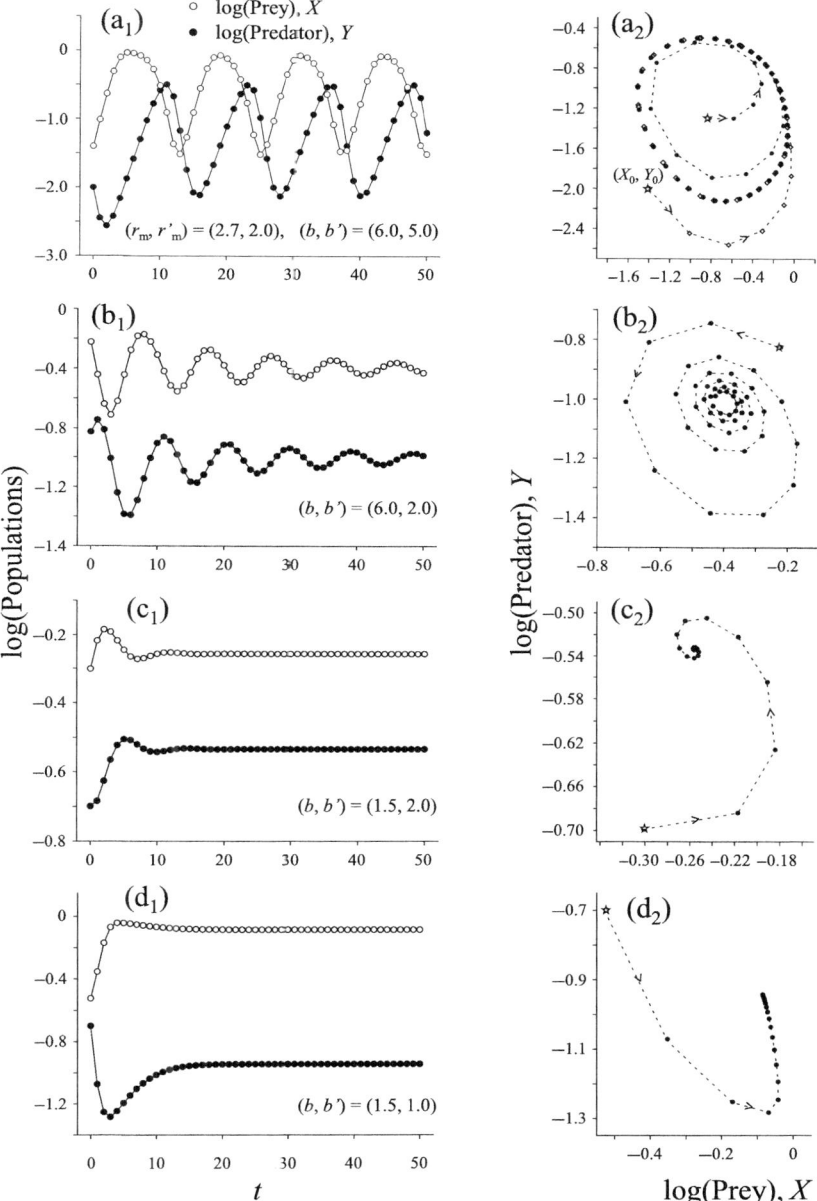

Figure 8.2 Examples of dynamical pattern in population that the system models (8.5) generate with different combinations of parameter values (r_m, r'_m, b, b').

192 · Predator–Prey Interaction Processes

$r'_m) = (2.7, 2.0)$, compared with $(6.0, 2.0)$ in the lynx example. The reduced skewness results in the lack of a 'dimple' in the phase-space representation in Figure 8.2a$_2$, compared with its counterpart in Figure 8.1d. Also explicitly shown in Figure 8.2a$_2$ is that, no matter where the initial state (X_0, Y_0), marked with a 'star', is placed, the phase-space track converges to the fixed elliptic orbit.

In Figure 8.2b, I reduced b' to as low as 2.0, while keeping $b = 6.0$, to indicate that the predator is not as efficient a hunter, while just as demanding as before. This situation yields damped oscillations for the following reason. The predator's high demand for food tends to periodically deplete the prey population to a sufficiently low level, and this keeps the system oscillating. Nonetheless, owing to its lower efficiency, the predator is unable to keep the cycles oscillating, resulting in damped oscillations. Yet, the hunting efficiency is high enough to make the speed of damping rather slow. In fact, the phase-space track in Figure 8.2b$_1$ converges to the equilibrium only after swirling a countless number of times beyond the capability of my computer software.

In Figure 8.2c, I set $b = 1.5$, much reduced from 6.0 in the foregoing examples, indicating that the predator demands much less. This causes the oscillations to damp down fast: its phase-space track in Figure 8.2c$_2$ swirls (spirals) only once in the example (in general it could swirl more than once, albeit only limited times) before practically converging to the equilibrium point. In Figure 8.2d, the predator's efficiency (b') is further reduced to 1, resulting in convergence with only a minimal sign of swirling. [Note: We are working on the dynamics of endogenous processes. Under exogenous influences, cycles would not damp down as will be shown in due course. Incidentally, Appendix 8B shows how to find the numeric values of the equilibrium levels for the prey X and predator Y of the interaction model (8.4).]

To summarize, we see that the predator's high demand and hunting skills would result in a substantial depletion of the prey population such that it requires several generations to recover. This in turn results in the so-called 'population cycles'; see [Note] below. In contrast, the predator's low demand for food and hunting skills, as typically exemplified in Figure 8.2d, implies that the prey is largely free from predation, a situation which in turn allows the predator to find a sufficient amount of food all the time. Thus, each species tends to retain its own single-species characteristic of dynamical pattern: in particular, each exhibits a nearly monotonic convergence (as in Figure 4.6a) because the parameter values (r_m, r'_m) are set as low as $(2.7, 2.0)$.

[Note: In ecology the term 'cycle' is commonly used to describe a tendency that increases for a few generations, followed by a decrease for another few generations. It may exclude a very short-term cycle, like a saw-toothed oscillation. However, in a broader sense, a cycle is a periodic repetition of any length. There are pros and cons of the common usage of the term 'cycle' from a semantics point of view. Thus, if necessary, we may distinguish the common usage as 'a cycle in the narrow sense'.]

8.4.2 A Few More Varieties

The dynamical patterns exhibited by the examples in Figure 8.2 are straightforward to comprehend in terms of the combinations of parameter values. The results of some other combinations may not be so straightforward. Figure 8.3 shows three examples that model (8.4), or formulae (8.5), generate with the combinations of parameter values (r_m, r'_m) and (b, b'), given in each graph. All graphs start with the same initial densities (x_0, y_0) ≈ (1.00, 0.063), or their $\log_{10}$ values (X_0, Y_0) = (0, −1.2). I now show how to interpret the pattern in each graph.

Example (Figure 8.3a). This is a variant of Figure 8.2c. The point of interest here is to find the mechanism underlying the saw-toothed oscillations in the initial section of the prey series $\{X_t\}$, which converges to a level by about $t = 10$. To explain this feature, let me rewrite model (8.4) compactly as:

$$r_t(x_t, y_t) = [r_m p(y_t)] \exp(-x_t); p(y_t) = \exp(-by_t) \quad (8.6a)$$

$$r'_t(y_t, x_t) = [r'_m q(x_t)] \exp(-y_t); q(x_t) = [1 - \exp(-b'x_t)]. \quad (8.6b)$$

We see that each process is of the same form as (4.1) for a single-species process, but the potential reproductive rate r_m is now associated with p (the effect of predation as a function of y_t), and the rate r'_m is associated with q (the effect of the availability of prey as a function of x_t). In other words, the potential reproductive rate of either species is no longer a fixed parameter but changes in association with changes in y_t or x_t.

In the present example, $Y_0 = \log_{10}(y_0) = -1.2$ (or $y_0 ≈ 0.063$), indicating a near absence of predation initially. So, with $b = 1.5$ (shown in the graph), we find $by_0 ≈ 0.095$ and $p(y_0) ≈ 0.9$. Multiplying this value by $r_m = 9.0$, we find $[r_m p(y_0)] ≈ 9.0 \times 0.9 = 8.1$ and, after an ln-transformation, $\ln[r_m p(y_0)] ≈ \ln(8.1) = 2.1$. This suggests that, for an initial few generations, and in the near absence of predation, the prey

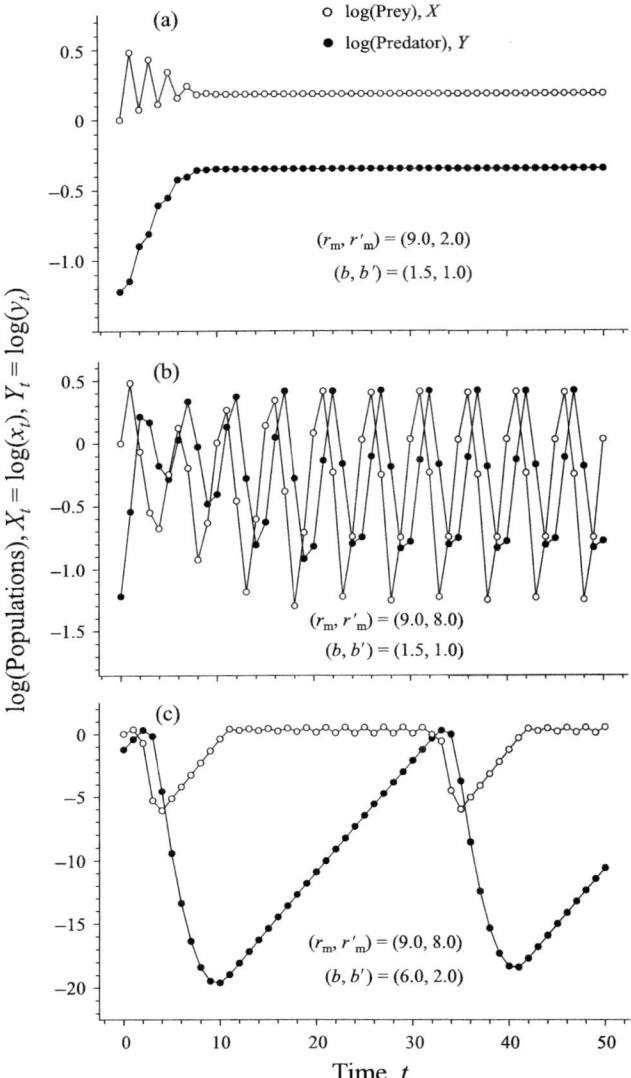

Figure 8.3 A few more examples of dynamical patterns that system model (8.5) generates.

population series $\{X_t\}$ exhibits a dynamical pattern very similar to one in Figure 4.6c (in which $R_m = 2.0$), i.e. a saw-toothed oscillation.

In the meantime, enjoying a good supply of the prey X, the predator series $\{Y_t\}$ increases rapidly to about -0.35 by $t = 10$, or $y_{10} \approx 0.447$.

8.4 Variation in Dynamical Patterns

Thus, a similar calculation as above results in $\ln[r_m p(y_{10})] \approx 1.527$, which is less than $R_m = 1.9$ in Figure 4.6b, a damped oscillation to the equilibrium level in the series $\{X_t\}$ for $t \geq 10$. Furthermore, as the series $\{X_t\}$ converges to the equilibrium level, the predator series $\{Y_t\}$ in turn converges to its equilibrium level as well. [Note: Evidently, in a two-species interaction process, neither species can converge to an equilibrium state on its own.] Altogether, we understand the ecological mechanism underlying this example. The following two examples have some problems regarding their ecological feasibility.

Example (Figure 8.3b). This is one of the limit cycles that the interaction model (8.6) generates, like one in Figure 8.2a. However, the particular combination of the parameter values here, i.e. $(r_m, r'_m) = (9.0, 8.0)$ and $(b, b') = (1.5, 1.0)$, makes the length of a cycle much shorter. At a glance, nothing appears to be particularly unusual with this pattern of population fluctuation. However, what I am concerned about here is the question: Is the combination $(r'_m, b') = (8.0, 1.0)$ feasible ecologically? In my perception, it sounds doubtful that a predator with poor hunting skills, as indicated by the low value of $b' = 1.0$, has a high potential reproductive rate $(r'_m = 8.0)$, an attribute that is unlikely to evolve. In other words, this example is mathematically feasible but could be unrealistic ecologically. The example in Figure 8.3c has a similar problem as discussed in the next paragraph.

Example (Figure 8.3c). The parameter values $(r_m, r'_m) = (9.0, 8.0)$ and $(b, b') = (6.0, 2.0)$ generate a rather weird dynamical pattern. Although its hunting skill is rather low $(b' = 2.0)$, the predator's high demand for food $(b = 6.0)$ depletes the prey density (X_t) substantially after a few initial generations. This in turn causes the predator population (Y_t) to plummet because its high demand for food, combined with mediocre skill, cannot sustain its population on too low a food supply. As Y_t has plummeted to an extremely low level, it takes so many generations (more than 20) to recover itself to the initial level, despite its high potential reproductive rate $(r'_m = 8.0)$. In the meantime, enjoying the near absence of predation, prey X maintains its own population in the manner dictated by model (4.1) for a single-species process (saw-toothed oscillations in this example), until the predator population Y_t recovers to the original level. But, then, the prey population X_t rapidly decreases again and Y_t accordingly plummets to repeat a new cycle, and so on and on goes the weird dynamics.

The question is: would we actually see this sort of dynamics occur in nature? Probably not. I can think of two reasons. One is that, as in the preceding example Figure 8.3b, the combination of low hunting skill ($b' = 2$) with high potential reproductive rate ($r'_m = 8$) is unlikely to evolve. The other reason is physical, as follows.

A close look at the predator series in Figure 8.3c, i.e. $\{Y_t = \log_{10}(y_t)\}$, reveals that the difference between its peak and its trough is as much as -20 units of scale on the vertical axis. This means that the Y series is depleted by as much as 10^{-20}-fold of its initial level y_0, or the difference is as much as billion × billion × 100-fold. The largest difference I have ever seen during my research career, and through my own eye, was in the magnitude of 100,000-fold (or maybe a bit more) in the spruce budworm (*Choristoneura fumiferana*) populations in New Brunswick, Canada. [Digression: Working in the woods at its outbreak level, you can feel and hear their frass falling like rain. Cut a branch and shake it, you see literally hundreds of the larvae fall off it. When scarce, more than a thousand branches collected across the entire areas of the province yield just a few dozen larvae altogether. You can only know that the budworm has not been extinct when you set up a pheromone traps: quite a few male moths (and hence females, too) are still around in many places.]

I have no idea what the situation could be like with a population of microorganisms, as it is outside of my expertise, but, as far as those animals that we usually deal with in population ecology are concerned, a 10^{-20}-fold decrease in density (from a usual level) implies, in all likelihood, the extinction of the population. That is, it is mathematically feasible, but is, so far as I perceive, doubtful to occur in nature.

So far, our investigations have been based specifically on the interaction model processes (8.1) and its variant forms (8.4–8.6). However, these are in turn based on the particular (single-species) model process (4.1). So, a generalization may be of interest.

8.4.3 Generalization of Predator–Prey Interaction Processes

Let us write model (8.1) in the general form:

$$r_t(x_t, y_t) = [r_m f(x_t)] p(y_t) \tag{8.7a}$$

$$r'_t(y_t, x_t) = [r'_m g(y_t)] q(x_t) \tag{8.7b}$$

in which $[r_m f(x_t)]$ and $[r'_m g(y_t)]$ represent the single-species processes of the prey and predator species, respectively: these are based on model (4.1) in (8.1) as mentioned. Now that (4.1) has been generalized to (5.12) in Chapter 5, why not use (5.12) as a basis for f and g to generalize interaction model processes?

Certainly, the generalization implies flexibility that could be useful (or even required, as will be touched upon later) in studying varieties of the interaction processes, either theoretically or observationally. However, as far as I have conceived, I do not see that the majority of natural processes are drastically different in structure from the general form (8.7). From this point of view, a further generalization in detail would not much enrich our knowledge about the principles underlying the predator–prey interaction processes that we usually encounter in nature. Besides, a generalization often results in algebraic complications that would likely hamper the way we conceive the principles. So, leaving the task of generalization to some of you who are interested in this subject, I proceed to talk about another subject that is fundamentally important in studying natural populations: the effects of random exogenous influences.

8.5 Effects of Random Exogenous Influences

So far, I have been dealing with endogenous mechanisms as a theoretically important subject for understanding the principal aspects of system dynamics. However, we would never see these endogenous processes per se in natural environments. What we actually see is those under exogenous influences which tend to hide the endogenous mechanism. So, ultimately, we wish to find a way to guess the hidden endogenous mechanisms underlying observed series. This is not easy, and may even be impossible, but without trying, we would never understand what we actually observe. To begin, let us do some simulations to get some basic ideas about the effect of random influences.

8.5.1 Preparing for Simulations

Notice that the predator–prey processes have, at least, four parameters (r_m, b) and (r'_m, b'), as in model (8.6), and that each of these parameters can be subject to exogenous influences. Hence, in general, there would be four independent ways to incorporate the effects of influences. However, taking all these effects into consideration at once would be too cumbersome algebraically. To get around the problem, I suggest the following simplification.

198 · Predator–Prey Interaction Processes

To begin, notice in (8.6) that parameter b controls the function $p(y_t)$, i.e. the effect of predator Y on prey X, and b' controls $q(x_t)$, i.e. Y's demand for food (X). However, p acts as a part of r_m, and q a part of r'_m. In other words, the exogenous influences that act directly on (r_m, r'_m) may also indirectly act on (b, b'). Thus, for simplicity, we may consider that only the parameters (r_m, r'_m) are subject to the effects of exogenous influences. The following is the simplest way to incorporate the effects into each of the processes for X and Y.

Letting (as before) an upper-case letter be the logarithm of the corresponding lower-case letter, and letting ε_t and ζ_t be the effects of exogenous influences on R_m and R'_m, respectively, we have the stochastic versions of the processes in (8.7) after ln-transformations:

$$R_t(X_t, Y_t) = (R_m + \varepsilon_t) + P(Y_t) + F(X_t) \qquad (8.8a)$$

$$R'_t(Y_t, X_t) = (R'_m + \zeta_t) + Q(X_t) + G(Y_t). \qquad (8.8b)$$

Furthermore for simplicity, let us assume that the exogenous effect ε_t or ζ_t is a series of independent, identically distributed (iid) random numbers. Here, using a computer random number generator, I generated two iid series for $\{\varepsilon_t\}$ and $\{\zeta_t\}$, both being normally distributed with the mean 0 and the (arbitrarily chosen) variance σ^2. But we must also decide on how the two series are related, as several situations are ecologically conceivable: $\{\varepsilon_t\}$ and $\{\zeta_t\}$ are (i) identical to each other; (ii) completely independent of (uncorrelated with) each other; or (iii) are (positively) correlated. As the predator and prey species usually occur in the same habitat space, it is natural to consider that they are under similar (if not identical) environmental influences. Hence, situation (iii) is most likely. Thus, we want to generate a pair of mutually correlated series. [Note: Elements of ε_t and ζ_t are iid such that ε_i (or ζ_i) is uncorrelated with ε_j (or ζ_j) for $i \neq j$, but here we attempt to correlate ε_i with ζ_i.] How can we accomplish this?

Suppose we want the two series to be correlated with the (expected) coefficient α, say. The following is a simple way to do it. We first generate two independent (uncorrelated) series, say, $\{u_i\}$ and $\{v_i\}$. Second, we multiply each series by the constants α and β that satisfy the relationship $\alpha^2 + \beta^2 = 1$, or $\beta = (1 - \alpha^2)^{1/2}$, using the positive solution only. Third, we create a new series $\{w_i\} = \{\alpha u_i + \beta v_i\}$. Then, series $\{w_i\}$, albeit an iid on its own, is expected to be correlated with series $\{u_i\}$ as much as α, and correlated with series $\{v_i\}$ as much as β. [I suggest that you verify this yourself: if you cannot, see Appendix 8C.] So, you can use the $\{w_i\}$ paired with either $\{u_i\}$ or $\{v_i\}$. Then, equating ε_i to u_i (or to v_i if you like) and ζ_i to w_i, we can incorporate them into (8.8). Further, writing the ln-transformed functions F, G, P, and Q explicitly (see [Note:] below), we find (8.8) to be:

$$R_t(X_t, Y_t) = (R_m + \varepsilon_t) - b\exp(Y_t) - \exp(X_t) \qquad (8.9a)$$

$$R'_t(Y_t, X_t) = (R'_m + \zeta_t) + \ln\{1 - \exp[-b'\exp(X_t)]\} - \exp(Y_t). \qquad (8.9b)$$

[Note: $P(Y_t) = \ln[\exp(-by_t)] = -b\exp(Y_t)$; $Q(X_t) = \ln[1 - \exp(-b'x_t)] = \ln\{1 - \exp[-b'\exp(X_t)]\}$; $F(X_t) = -\exp(X_t)$; and $G(Y_t) = -\exp(Y_t)$. Substituting these in (8.8), we have (8.9).] These expressions are a little cumbersome but are a necessary evil for maintaining the notational consistency (after the ln-transformations) for doing numerical computations with a computer as well as for graphics. So bear with them.

Now let us apply processes (8.9) to the four pairs of endogenous series in Figure 8.2.

8.5.2 Simulations

Let us use the series $\{\varepsilon_t\}$ and $\{\zeta_t\}$ that are 80% correlated, i.e. $\alpha = 0.8$ in the foregoing method.

[Why 80%? I simply chose it because if $\alpha = 0.8$ exactly, then $\beta = 0.6$ exactly, because $0.8^2 + 0.6^2 = 1$ exactly. That is, these are the easiest numbers that I happen to use in the simulations.] Also, the variance of each series is conveniently set equal to 0.01 such that we see the effect of exogenous influences clearly: too low or too high a variance tends to obscure the effect.

Figure 8.4 shows four pairs of stochastic series, based on their respective endogenous versions in Figure 8.2, although plotted in a $\log_{10}$-scale. In addition, the exogenous influences $\{\varepsilon_t\}$ and $\{\zeta_t\}$ are shown in the bottom graphs. In the following, each pair of series will be referred to by the respective alphabetical code: pair (a, b, c, and d).

Pairs (a) and (b): We recognize in appearance that pair (a) retains the limit-cycle characteristics of its endogenous counterpart in Figure 8.2a, but the amplitude and the periodicity of a cycle vary in haphazard manners. Also recognizable is that each series of pair (b) retains the cyclic trend of its endogenous counterpart in Figure 8.2b. However, unlike the counterpart, the stochastic pair shows no sign of damping down. This is

200 · **Predator–Prey Interaction Processes**

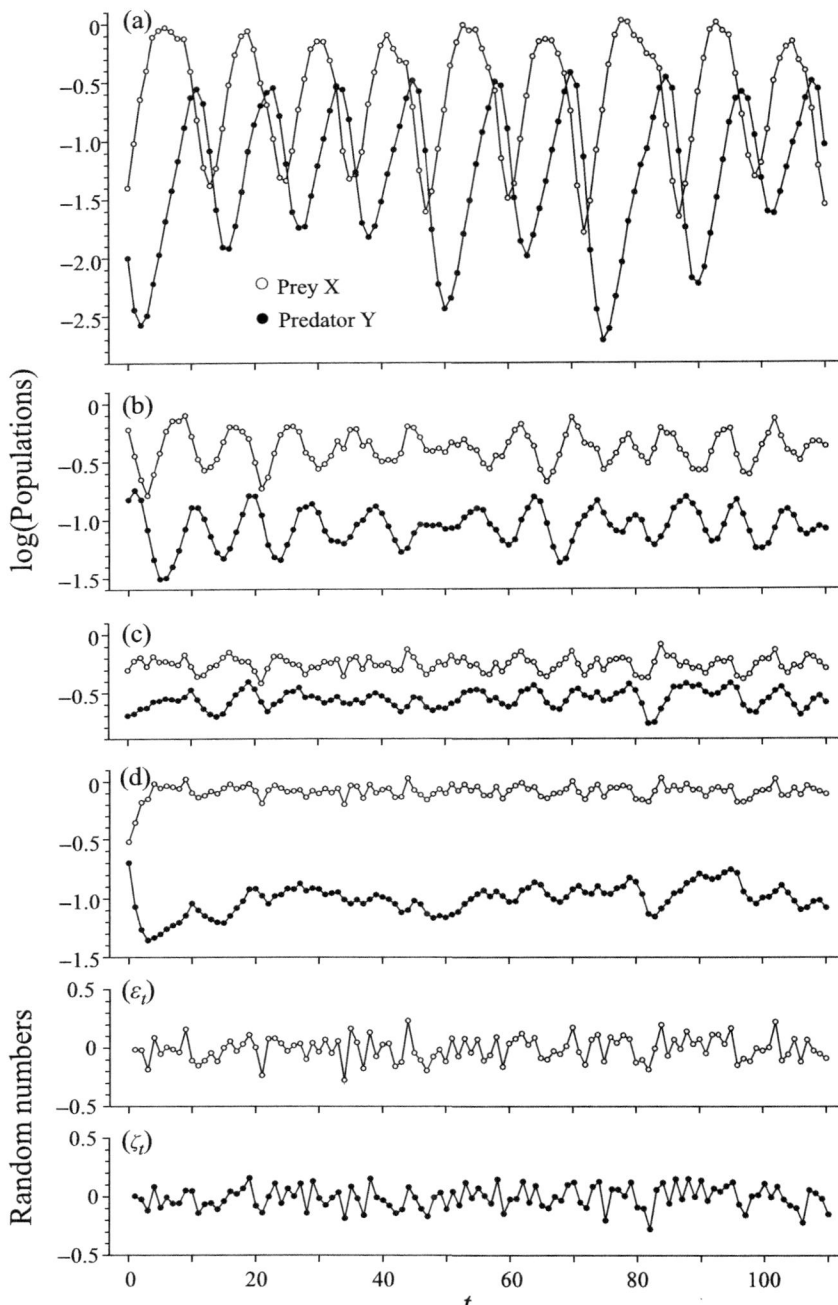

Figure 8.4 (a–d) Same as the endogenous series of Figure 8.2a$_1$–d$_1$, respectively, but subjected to the (iid) random exogenous influences of the bottom two graphs.

because the series are continually perturbed (agitated) by the random exogenous influences, and are always kept away from converging to the equilibrium level.

Pair (c): Because of the constant agitations by the exogenous influences, the series in this pair do not converge to their equilibria but keep fluctuating. What is more, a close look at the series reveals a sign of their own endogenous cyclic trend that recurs every so often, if not always. In fact, this attribute is a manifestation of the corresponding (endogenous) phase-space orbit in Figure $8.2c_2$ exhibiting a tendency to swirl, even though converging quickly to the equilibrium point.

Pair (d): The endogenous counterpart of this pair (Figure $8.2d_1$) converges to equilibria even more quickly than in pair (c). This endogenous attribute (little tendency to cycle) is manifest in the X series (open circle) being highly correlated (as much as 93% in fact) with the non-cyclic (iid) series $\{\varepsilon_t\}$. However, somehow, the Y series (solid circle) does not behave like the (paired) X series: it is not as highly correlated (merely by 33%) with series $\{\zeta_t\}$. The reason for this will be discussed shortly.

The above simulations show how exogenous influences affect the endogenous attributes of the interaction processes. I now suggest that we venture into a challenging task of bringing out the endogenous process, buried under the noisy exogenous influences in an observed population series. The method of autocorrelation function (ACF), which I briefly referred to in Chapter 7, may work to an extent. So, to begin, let us take a look at it.

8.5.3 Utilization of the ACFs

An autocorrelation is measured by the usual (Pearson) correlation coefficient between the pair of data points (in an observed series) that are so many time steps apart from each other. If we correlate pairs of adjacent points in a series of length t, like (1st, 2nd), (2nd, 3rd), (3rd, 4th) ... ($t-1$-st, t-th), it is called the autocorrelation for lag 1. Likewise, correlation between two points, skipping a few points in between, like (1, 5), (2, 6), (3, 7), ..., ($t-4$, t), is the autocorrelation for lag 4, and so on. Then, the set of autocorrelations as a function of the lag l (= 1, 2, ...) is called an autocorrelation function (ACF). In particular, a graphical plot of the calculated autocorrelation coefficients against their lag l is called the 'correlogram'. [Digression: Legend has it that the late Professor Shunro Utida of Kyoto University hired a few female students to hand-calculate the ACFs with his experimental series of bean weevils. It took them a week or more to finish the tedious job. Why girls? Apparently, they were more meticulous and dependable than boys in doing tedious work. Nowadays, a statistical package in computer software does the job in an instant.]

There are various uses of the ACFs in statistics but, in the present context, we look into a correlogram to extract an endogenous pattern buried under the unwanted environmental noise. [Note: A calculated correlogram is usually accompanied with a pair of curves to indicate a 95% confidence interval for no correlations (e.g. as in an iid series). In the present context, we pay particular attention to the pattern of the correlogram, disregarding the confidence interval. After all, 'insignificantly different from zero' does not imply 'no correlation': it is simply difficult to distinguish from 0 with a certain level of accuracy.]

Figure 8.5(a–d) shows the calculated correlograms of the respective stochastic series in Figure 8.4(a–d), which in turn correspond to the endogenous series in Figure 8.2(a_1–d_1). [Note: It is customary to use bar graphs to represent a correlogram. It does make good sense.] In calculating the ACFs, I use only the section of each series after it has apparently reached a steady (stationary) level, discarding the initial section for $t < 10$. [Note: The reason for using the stationary section of a series will be given shortly. For the concept of stationarity, see in Chapter 7.] The resultant correlograms for prey and predator series are paired in each alphabetically designated graph. The following is what we see in detail.

The pair of correlograms in (a) reveals two major attributes of the corresponding endogenous cycles in Figure 8.4a. First, the correlograms exhibit a peak-to-peak (or trough-to-trough) cycle of the period 12 generations. Second, the endogenous limit cycle is manifest in the continually oscillating correlograms with a very slow rate of damping down (see Note below). Likewise, the correlograms (b) exhibit an oscillatory tendency (of the period 10–9 generations) but damp down more quickly than those in (a), revealing their endogenous tendency (Figure 8.2b) of convergence to equilibria. Even the correlograms (c) of the rather fuzzy series in Figure 8.4c clearly exhibit a cyclic tendency, although in a reduced amplitude (lower correlation coefficients), indicating the fuzziness of the observed series.

[Note: Most correlograms damp down eventually because, by virtue of the random nature of the series, an autocorrelation diminishes as the lag increases. Evidently, while what happens today could be expected to be well correlated with what happened yesterday, we would not expect what happens 100 years from now to be as well correlated. Thus, even for a limit-cycle series, when becoming a stochastic process under

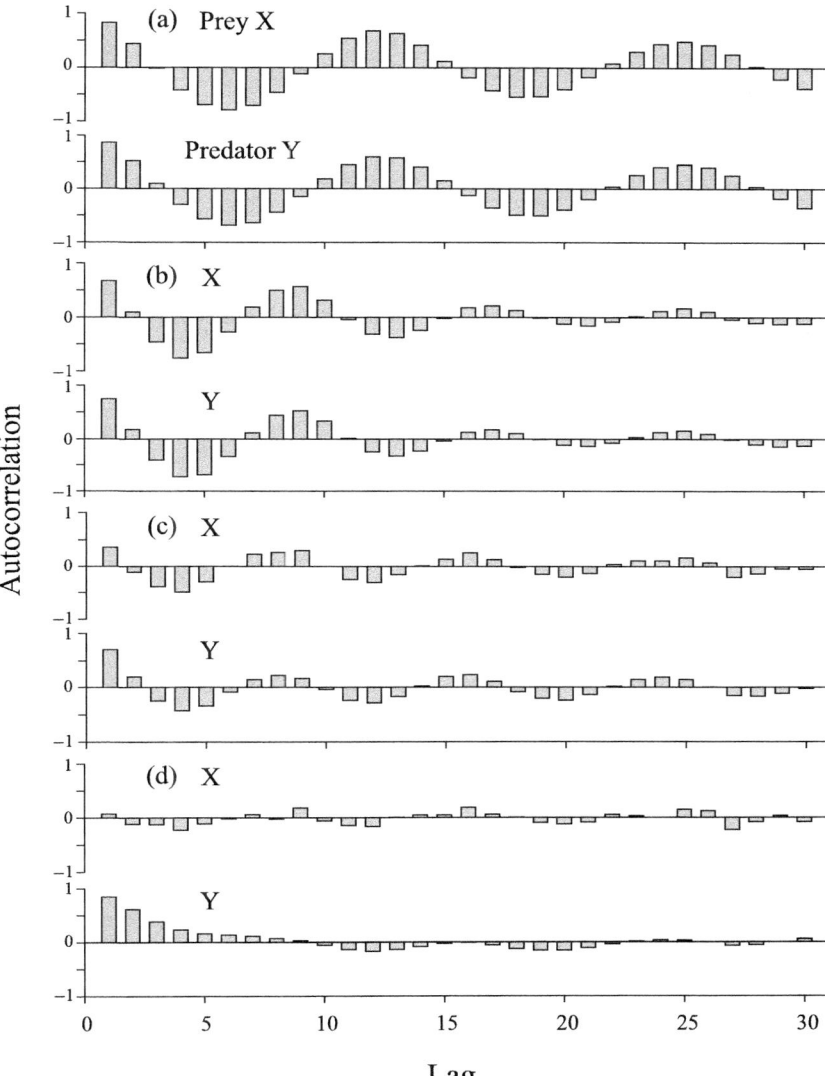

Figure 8.5 Autocorrelation functions (or correlograms) of the simulated series in Figure 8.4.

random exogenous influences, its correlogram would exhibit a tendency to damp down as the lag increases, but only slowly. Obviously, the higher the variance of the exogenous influences, the quicker the rate of damping would be.]

204 · Predator–Prey Interaction Processes

As for the pair of series (d) in Figure 8.4, let us look at the prey X series first. We see that the autocorrelations for all lags are low (in absolute value), so low in fact that they would usually be considered to be insignificantly different from zero. Yet, we still see a cyclic pattern as a manifestation of the swirling tendency of its endogenous counterpart in its phase-space representation in Figure $8.2d_2$. Now, take a look at the correlogram for the Y (predator) series in Figure 8.5d. It appears quite different from the correlogram for its counterpart, the X (prey) series. Why does it differ so much? In fact, this is known as the problem of a nonstationary series. This is an important issue in dealing with observed population series, an issue not well recognized in the ecological literature. So, it should be explained in some details.

8.5.4 Correlogram of a Nonstationary Series

As it so happens by chance (not so rare an occurrence in stochastic processes), the Y series in Figure 8.4d exhibits a temporary trend (for the first 100 or so generations): I copy it in Figure 8.6a in which the original graph (in Figure 8.4d) is vertically expanded to bring out the trend. Now, as is well known in statistics, a correlogram is sensitive to a trend in the series (or generally speaking, sensitive to non-stationarity) and may distort the endogenous aspects of the series we are trying to extract. [Note: If certain statistical moments of a series, usually its mean and covariances, are time-invariant, the series is said to be 'stationary': see Section 7.9.1 in Chapter 7; conversely, it is said to be 'non-stationary' if these moments vary with time.] Certainly, the graphed section of the Y series in (d) is non-stationary as its mean is increasing with time, and its correlogram does not reveal the cyclicity that the series in fact exhibits about its trend. But, is there a remedy? Yes, detrending, i.e. to get rid of the trend.

You may detrend a series by calculating the deviations of data points about the (estimated) trend line. The resultant series of deviations should be trendless. However, a method often used in statistics (quicker, easier, yet effective) is 'differencing' the original (non-stationary) series. [Note: Differencing is to calculate the difference between the adjacent data points, e.g. $Y_{i+1} - Y_i$ for $i = 1, 2, \ldots, t$. Again, your computer software (statistical packages) can do it instantly.] It so happens in our present case that the difference has a particular ecological meaning: it is equal to the log rate of change (R'_t) in the predator density Y_t from t-th to $t + 1$st generations, i.e. differencing estimates $R'_t = Y_{t+1} - Y_t$. The result is shown in Figure 8.6b, and its correlogram (after discarding the initial,

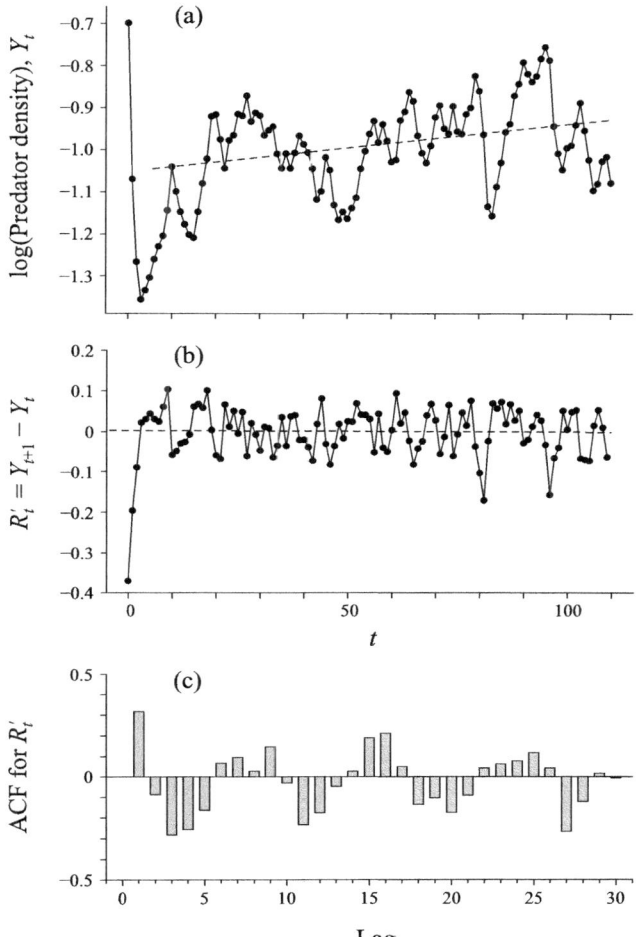

Figure 8.6 Revealing the cyclic trend in the predator series of Figure 8.4d.

non-stationary 10 data points) is given in Figure 8.6c, which clearly shows a cyclic tendency (of period 8 or so). It suggests that the Y series here is indeed cyclic, but about its removed trend, exhibiting the periodicity similar to that of its paired X (prey) series.

Altogether, we see that a correlogram can to an extent detect the dynamical pattern of an endogenous series, buried under the fuzzy exogenous influences. However, this is just about as much as the correlograms can reveal. Naturally, though, we want to know more about the hidden aspects of the endogenous process. So, I suggest that we go one more step forward. But, for a preparatory step, I suggest that we construct endogenous reproduction surfaces as a generalization of the reproduction curves introduced in Chapter 4.

8.6 Reproduction Surfaces of a Predator–Prey Process

Recall that, as introduced in Chapter 4, a reproduction curve is formed by plotting the (endogenous) rate of change r_t ($= x_{t+1}/x_t$) against x_t in a single-species population process. The curve is further log-transformed to be used for graphical analyses of the population process. For instance, Figure 4.3a and its variants Figures 4.5a and 4.7 illustrate how a variety of population series can be generated by endogenous process (4.1) in accordance with changes in value of the parameters involved. In principle, the same method can be applied to the predator–prey interaction processes. However, unlike a single-species process, which is represented by a curve in a two-dimensional (2D) coordinate space, a two-species interaction process is represented by a pair of reproduction surfaces (one for each species) in a three-dimensional (3D) space. In particular, for prey X, its reproductive rate $r_t(x_t, y_t)$ is plotted against x_t and y_t; and for predator Y, the rate $r'_t(y_t, x_t)$ is plotted against y_t and x_t.

Figure 8.7 illustrates these reproduction surfaces, based on the system model (8.6) with the set of parameter values I conveniently chose for the illustration purpose. The vertical axis represents the reproductive rate $r_t(x_t, y_t)$ in Surface X, and the rate $r'_t(y_t, x_t)$ in Surface Y. The two horizontal axes in each surface represent the population densities x and y, the arrow at one end of each axis indicating the direction of increase in value.

However, in an attempt to make use of these reproduction surfaces as an analytical device for studying a predator–prey process, we encounter a technical problem: these three-dimensional surfaces per se are awkward for graphical analyses. To get around the problem, I extract (from a 3D surface) what I call the 'conditional reproduction curves' in a 2D graph as in a single-species process.

8.6.1 Conditional Reproduction Curves

Let us go back to the (endogenous) system model (8.6), and consider that we temporarily fix the predator density y_t in (8.6a) at a given value, i.e. we treat y_t as a constant. Then, we consider the rate $r_t(x_t; y_t =$ given constant) in (8.6a), which is compactly written $r_t(x_t | y_t)$. Likewise, we fix

8.6 Reproduction Surfaces · 207

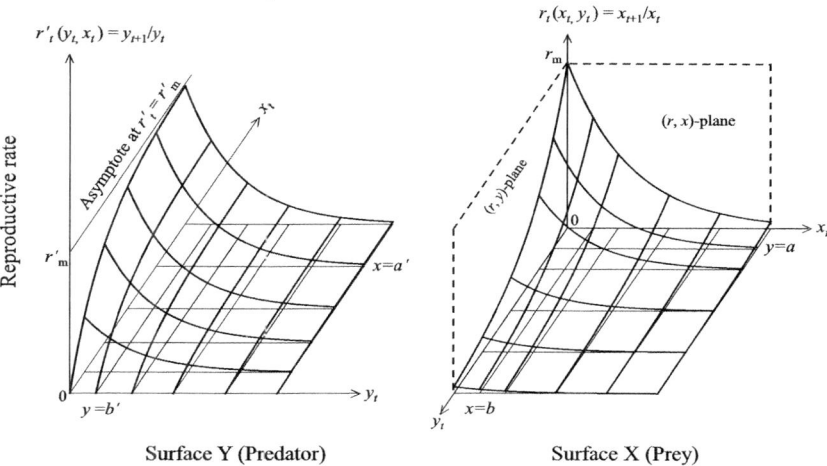

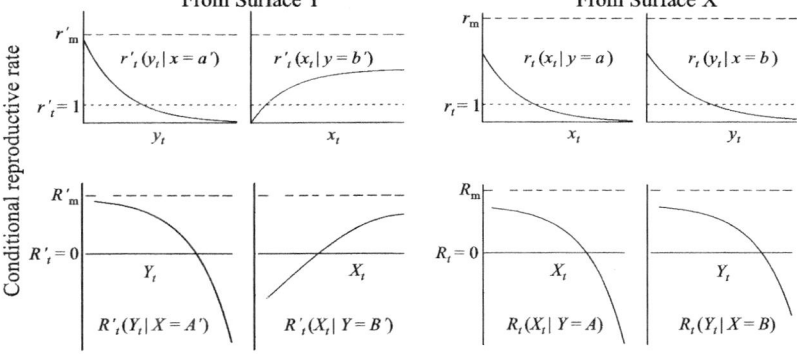

Figure 8.7 Reproduction surfaces of prey species X and predator species Y, and their corresponding conditional reproduction curves.

x_t in (8.6b) and consider the rate $r'(y_t; x_t =$ given constant), written $r'_t(y_t | x_t)$. The vertical line reads 'given' or 'conditional on', and hence I call $r_t(x_t | y_t)$ and $r'_t(y_t | x_t)$ 'conditional reproductive rates', written fully as:

$$r_t(x_t | y_t) = [r_m p(y_t = \text{given constant})] \exp(-x_t) \tag{8.10a}$$

$$r'_t(y_t | x_t) = [r'_m q(x_t = \text{given constant})] \exp(-y_t). \tag{8.10b}$$

We see that $r_t(x_t | y_t)$ is a function of x_t only, and $r'_t(y_t | x_t)$ a function of y_t only. Then, the plot of the conditional rate $r_t(x_t | y_t)$ in (8.10a) against x_t forms the 'conditional reproduction curve' for X (given y_t). Likewise, plotting $r'_t(y_t | x_t)$ in (8.10b) against y_t forms the conditional curve for Y (given x_t).

Furthermore, we can conceive another set of conditional reproductive rates in each of X and Y. In this set, we fix x_t in (8.6a) for X such that its conditional rate becomes $r_t(y_t | x_t)$, i.e. a function of y_t, given x_t = constant. Likewise, we fix y_t in (8.6b) for Y such that its conditional rate $r'_t(x_t | y_t)$ is a function of x_t, given y_t = constant. Thus, we have:

$$r_t(y_t | x_t) = r_m p(y_t)[\exp(-x_t = \text{given constant})] \quad (8.10c)$$

$$r'_t(x_t | y_t) = r'_m q(x_t)[\exp(-y_t = \text{given constant})], \quad (8.10d)$$

which form the conditional reproduction curves, each being paired with (8.10a) and (8.10b), respectively.

Now, let us go back to Figure 8.7. We see that each reproduction surface is graphed by means of the two sets of curves at some (arbitrary) intervals. For instance, in Surface X, one set of curves is drawn in parallel with the (r, x)-plane spaced at intervals along the y-axis, and the other in parallel with the (r, y)-plane spaced along the x_t-axis.

We see that each of these curves represents a conditional reproduction curve, given the (fixed) value of y or x. For instance, in Surface X, the curve in parallel with the (r, x)-plane marked $y = a$ is a conditional curve $r_t(x_t | y = a)$ plotted against x_t. Also, the curve in parallel with the (r, y)-plane marked $x = b$ is a conditional curve $r_t(y_t | x = b)$ plotted against y_t. These curves are shown right below Surface X. Likewise, the conditional curves $r'_t(y_t | x = d')$ and $r'_t(x_t | y = b')$ extracted from Surface Y are shown right below it. All of these conditional curves are transformed to logarithms in the respective bottom graphs, marked with the upper-case letters, e.g. $R_t(X_t | Y = A) \equiv \log[r_t(x_t | y = a)]$.

With these log-transformed conditional curves, we can readily visualize what the (log-transformed) surfaces X and Y would look like. [Note: I may use the expression 'log' indicating either 'ln' or '$\log_{10}$' unless specification is necessary.] Now the details.

8.6.2 Characteristics of Log-Transformed Reproduction Surfaces

The log-transformed conditional curves in Figure 8.7 have a graphically convenient feature. To see it, take a look at (8.7a) for X as an example. After the ln-transformation of the equation, we find:

$$R_t(X_t, Y_t) = [R_m + P(Y_t = \text{constant})] - \exp(X_t) \quad \text{(8.7a ln-trans)}$$

in which $R = \ln(r)$, $X = \ln(x)$, $Y = \ln(y)$, and $P = \ln(p)$. [Note: Here, I use the ln-transformation because it makes the algebraic expression simpler than the $\log_{10}$-transformation.]

Notice that (8.7a ln-trans) is identical in form to (4.4b) in Chapter 4, which has a convenient feature. As explained in Section 4.8.2 of Chapter 4, every curve generated by (4.4b) is of exactly the same shape regardless of the value of R_m; only its vertical position relative to the $(R_t = 0)$-axis is dictated by R_m. We see then that the expression $[R_m + P(Y_t = \text{const})]$ in (8.7a ln-trans) behaves in the same way as R_m in (4.4b) does. Thus, any conditional curve formed by (8.7a ln-trans) is of the same shape as (congruent with) other curves, regardless of the given (constant) value of Y_t. Only its vertical position, relative to the $(R_t = 0)$-axis, depends on the value of $[R_m + P(Y_t = \text{const})]$ whose upper limit is R_m because $P \leq 0$ as defined. This means that every vertical cross-section of the log-transformed Surface X (be it ln or $\log_{10}$), parallel to the (R_t, X_t)-plane, is identical in shape, differing only in height along the Y_t-axis. In other words, the log-transformed reproduction surface is represented by a family of the conditional curves in the identical shape, but their vertical positions decrease along the Y_t-axis as Y_t increases. The same applies to the cross-sections in parallel to the (R_t, Y_t)-plane. Surface Y is similarly visualized.

Based on these attributes of the reproduction surfaces, we can graphically generate a population series. The following example of the graphical method paves the way to our destination: estimation of the hidden endogenous features of an observed series. So, here we go.

8.6.3 Graphical Method of Generating a Population Series

Let us try to graphically generate the series in Figure 8.2a as an example. These series (before log-transformations) are generated by the formulae (8.5) with the parameter values and the initial densities: $(r_m, b, x_0) = (2.7, 6.0, 0.04)$ for the prey (X) series, and $(r'_m, b', y_0) = (2.0, 5.0, 0.01)$ for the predator (Y) series. Substituting these numeric values in (8.6), we can calculate the conditional reproduction curves $r_t(x_t | y_0)$ for the prey X, and $r'_t(y_t | x_0)$ for the predator Y in the following procedure.

After the $\log_{10}$-transformation for the graphics, $R_t(X_t | Y_0)$ is plotted against X_t to form the conditional curve, shown in the supplementary graph marked 'Prey X' (small graph, top left) in Figure 8.8: look for the

210 · **Predator–Prey Interaction Processes**

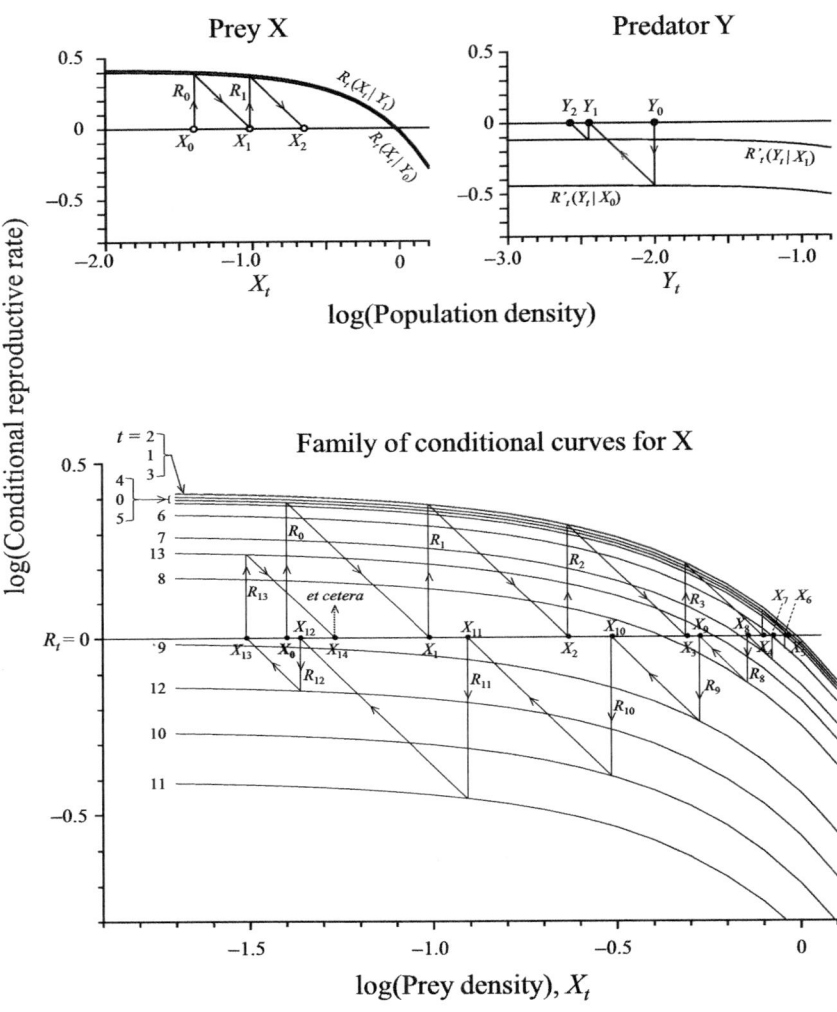

Figure 8.8 An example of graphically generating a population series by means of the family of conditional reproduction curves in a predator–prey interaction process.

lower curve marked $R_t(X_t \mid Y_0)$. [Note: This curve corresponds to the bottom graph in Figure 8.7 marked $R_t(X_t \mid Y = A)$ in which $A = Y_0$.] Then, we do the following (in the manner we did in Figure 4.3a): position X_0 on the horizontal ($R_t = 0$)-axis; from X_0, draw a vertical line upwards to hit the (lower) curve; draw a 45° line down to the right to find X_1. We see that $X_0 < X_1$, an increase in prey population from $t = 0$ to 1 as we actually see in Figure 8.2a (open–circle series).

Similarly, we calculate $r'_t(y_t | x_0)$ for the predator Y; $\log_{10}$-transform it; and plot it against Y_t to form the conditional curve $R'_t(Y_t | X_0)$, as shown in the supplementary graph ('Predator Y' on top right) in Figure 8.8. Then, mark Y_0 on the horizontal axis and find Y_1, following the arrowed lines [Note: This curve corresponds to the bottom graph in Figure 8.7 marked $R'_t(Y_t | X = A')$ in which $A' = X_0$. However, in the Figure 8.8 graph, the Y_t-axis is limited in the range $Y_t < -1.2$ where the curve is rather flat.] We see that, in this case, the conditional curve $R'_t(Y_t | X_0)$ is below the $R'_t = 0$, and hence $Y_1 < Y_0$, a decrease in predator population, as we actually see in Figure 8.2a (solid–circle series).

Now, we have a new pair of densities (X_1, Y_1). So, we calculate the conditional curves for $t = 1$, i.e. $R_t(X_t | Y_1)$ for X and $R'_t(Y_t | X_1)$ for Y, and draw the new curve in each of the (top two) supplementary graphs in Figure 8.8. [Note: We see that the second curves (for both X and Y) are above their respective first curves. The reason is as follows. Because $Y_1 < Y_0$ and $X_0 < X_1$, the second curves are further away from the front of their respective surfaces (in Figure 8.7) such that each one is higher than its respective first curve.] Repeating the same procedures (as with the first curve for each of X and Y), we find X_2 and Y_2. In particular, we see that $X_1 < X_2$ (a further increase in the prey population) and $Y_2 < Y_1$ (a further decrease in the predator population) from $t = 1$ to 2, as indeed seen in Figure 8.2a.

A further extension of the foregoing operations would generate a family of conditional curves for both species. The main (big) graph in Figure 8.8 shows the family of these conditional curves only for prey X (including the first two curves on the supplementary graph); the one for Y is not shown for the reason given in the [Note] below. On the left end of the main graph, the curves involved are marked $t = 0, 1, 2, \ldots$, indicating that the values of $(Y_0, Y_1, Y_2, \ldots)$ are given (fixed) with which the respective conditional curves have been calculated. Let us call each of them 'Curve(t)', $t = 0, 1, 2, \ldots$ [Note: The family of the curves for the predator Y to determine the given values $Y_0, Y_1, Y_2, \ldots$, is not created. This is because we only need the numerical values of $Y_0, Y_1, Y_2, \ldots$ to pair with $X_0, X_1, X_2, \ldots$ for calculating the respective conditional curves for X. Thus, as a short cut, I used the Y_t-values already calculated in drawing the Y-series in Figure 8.2a. After all, this graphical exercise is for understanding how a reproduction surface generates a population series. The actual generation of the series is done by numerical computation.]

212 · Predator–Prey Interaction Processes

Now, we are ready to graphically generate the series $\{X_t\}$ in the main graph in Figure 8.8. To begin with, look for the dot marked X_0 on the horizontal ($R_t = 0$)-axis near the left end of the graph. From the dot X_0, follow the upward vertical line as it hits Curve(0). Further, follow the 45° line to find X_1. Then, follow the vertical line (on the dot for X_1) as it hits Curve(1) in the middle of the uppermost three curves, though a little difficult to see. Further, follow the 45° line to find X_2. And so on and on, you see the dots marked X_3, X_4 also lined up towards the right until you come to the dot X_5 near the (right) end of the graph. Thus, you know how the first six points of the X-series (including X_0 in Figure 8.2a) are generated. Then, you see the point X_6 is above the corresponding Curve(6), or that section of the curve (corresponding to the point X_6) is below the horizontal ($R_t = 0$)-axis, indicating a decrease in population. Thenceforth, by following the vertical and 45° lines in the direction of arrows, you see that the subsequent series $\{X_t\}$ for $t = 6, 7, \ldots, 13$ decrease to complete the first cycle of the X-series in Figure 8.2a and a new cycle begins.

I now show that by reversing the foregoing procedures, we may graphically see the hidden endogenous conditional curve in a pair of observed (prey–predator) population series.

8.7 Revealing Conditional Reproduction Curve in Observed Series

8.7.1 Principle of the Method

Let us assume that the pair of simulated stochastic series X (prey) and Y (predator) in Figure 8.4a is an actually observed pair $\{X_t, Y_t\}$ in the field or in the laboratory. Our goal here is to graphically reveal the endogenous conditional reproduction curves (e.g. of Figure 8.8) that are (assumed to be) hidden in the observed series. To illustrate how we can achieve the goal, I select the prey species X (cf. Surface X in Figure 8.7) as an example.

To begin, using your imagination, visualize the observed pairs of points (X_i, Y_i), $i = 0, 1, 2, \ldots, t$, of Figure 8.4a (the stochastic version of Figure 8.2a), scattered about the log-transformed Surface X, some deviating above and others below the surface by as much as the random influences ε_i (cf. Figure 8.4 second from bottom). Now, consider that we have a bird's eye view of the surface X with the data points, as shown in Figure 8.9: that is, we have the vertical projection of the data points of the series X (in Figure 8.4a) onto the bottom plane of Surface X, i.e. the (X_t, Y_t) coordinate plane.

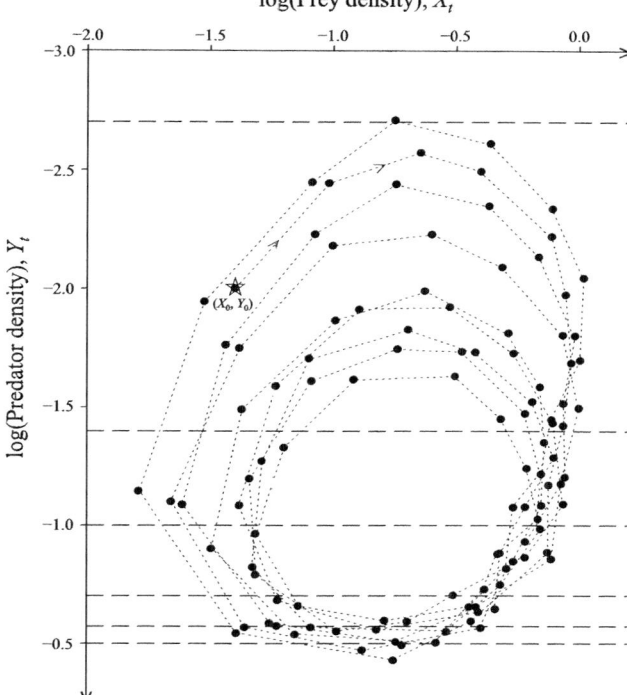

Figure 8.9 The vertical projection (bird's eye view) of the data points of the series $\{X_t\}$ in Figure 8.4 onto the bottom plane of Surface X, i.e. the (X_t, Y_t) coordinate plane in Figure 8.7 after a log-transformation.

What we see here is in fact a stochastic version of the endogenous phase-space representation in Figure 8.2a$_2$. [Note: In Figure 8.9, the direction of increase of the Y_t-axis is reversed for convenience so that the phase-space orbit goes clockwise.] In addition, the horizontal (dashed) lines, drawn in parallel to the X_t-axis at certain intervals, are the bird's eye view of the endogenous conditional reproduction curves (vertical cross-sections of Surface X), given the corresponding Y_t values on the left-hand axis. The width of each interval between the adjacent lines decreases in the direction of increasing Y_t (i.e. from top to bottom of the graph) to indicate the following attribute of the surface.

Visualize that Surface X is curved downwards with increasing steepness along the Y_t-axis in the increasing (arrowed) direction. In other words, the surface is flatter (or steeper) for lower (or higher) values of Y_t. Thus, the intervals between the dashed lines are so determined that the

214 · **Predator–Prey Interaction Processes**

height differentials between the adjacent vertical cross-sections are more or less equalized; the reason for the equalization will become apparent in a moment.

We now change our view (from the bird's eye to a frontal view) of Surface X as shown in Figure 8.10. What we see here is the horizontal projection of the data points onto the (R_t, X_t) plane (of Surface X). The dashed curves are the conditional curves, corresponding to the respective dashed straight lines in Figure 8.9. We see that these curves (in

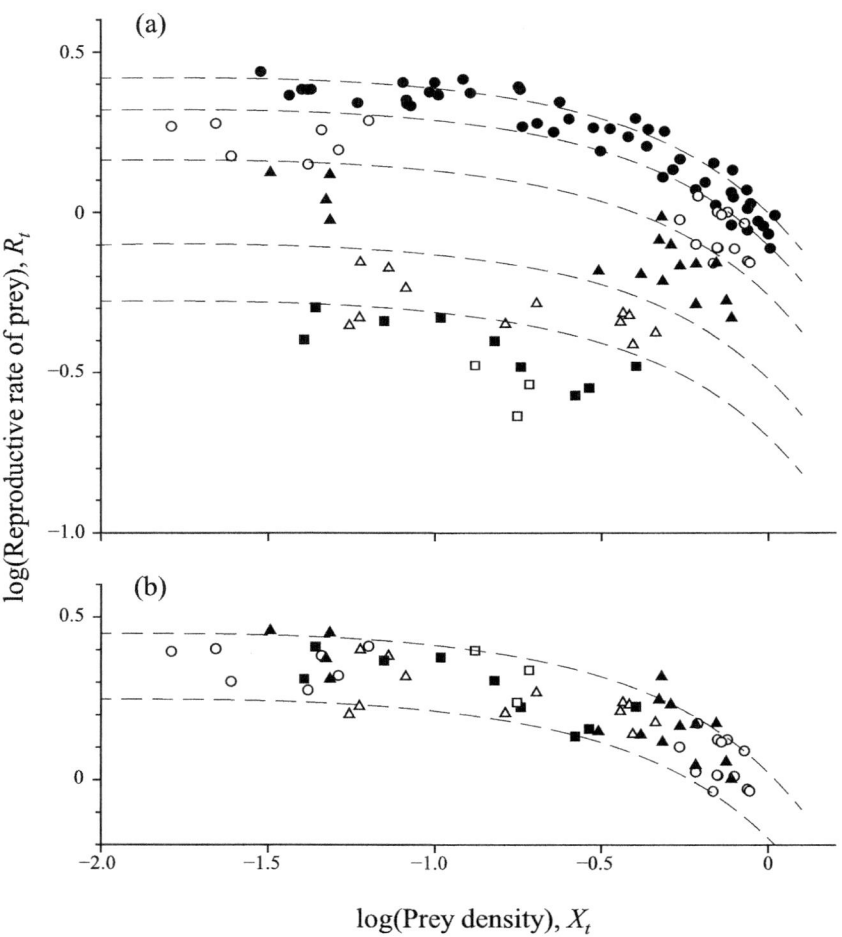

Figure 8.10 Same as Figure 8.9 but a frontal view, i.e. the horizontal projection of Surface X with the data points onto the (R_t, X_t) plane.

8.7 Revealing Conditional Reproduction Curve · 215

Figure 8.10) are now spaced with more or less equal widths. Now I show that these curves (supposed to have been hidden in the observation) can be revealed by means of the observed data points.

In Figure 8.10a, those data points that lie within a given interval in Figure 8.9 are marked with the same symbol. Thus, we see that the points of the same kind lie mostly within their respective interval (band) between the adjacent conditional curves. In other words, these points of the same kind (as a set) actually trace the shape of the (supposedly hidden) endogenous curve that we have attempted to reveal. Likewise, albeit not shown, the side view (perpendicular to the frontal view) of Surface X would give the projections of the data points onto the (R_t, Y_t) plane, which would trace the conditional curves, formed by plotting $R_t(Y_t | X_t)$ against Y_t. I leave this task for your homework.

8.7.2 Further Manipulation for a Better View of Conditional Reproduction Curves

In the Figure 8.10a example, only those data points within the first (top) interval trace the shape of the curves to an adequate length. In the second and third intervals, the data points occur only about the two extreme ends of their range of distribution, lacking anything in between. In the bottom intervals, they are aggregated only in the middle range of their distribution. However, if we equalize these intervals in height by stacking up one over the other at a more or less equal level, they would complement each other to form a continuous distribution, as shown in Figure 8.10b.

In the foregoing example, we knew the hidden endogenous curves which enabled us to divide the surface into appropriate intervals. The question is: how can we find the intervals in an actually observed series without knowing the endogenous curves? As far as I can see, there is no systematic way to find them but in a trial-and-error method. To do it, I suggest the following. You first divide the phase-space graph (as in Figure 8.9) into intervals by guessing. Then, look at the frontal view (as in Figure 8.10), and iterate the procedure until you see curves in a consistent manner. For practice, do a lot of simulations with models (8.9), pretending that you do not know its endogenous attributes. Have fun!

However, in doing practise simulations, consider the following. As already mentioned, the foregoing example is based on the basic model process (8.6) which is based on the (single-species) model process (4.1) which, in turn, is a particular case of the general model (5.12). So, for comparison with an actual observation, which would likely require flexibility, you should consider the use of (5.12) as a generalized model. Then, hopefully, one of them would match the observation to an acceptable degree.

8.7.3 Limitations

The foregoing method works well if your observations meet the following conditions: (i) there must be a paired set of data for both prey and predator; (ii) the data must cover an adequate section of the endogenous surface to be estimated; (iii) the observed process obeys the general structure (8.7 or 8.8). The Figure 8.4a example meets all three conditions as it is an artificially created process for the purpose. Unfortunately, I have seldom seen a field study that meets the first condition, let alone the second, and the possibility of testing the third is out of the question. The reality is that we rarely find suitable materials, situations, and (time or financial) opportunities for conducting good-enough field studies. Then, after having gone through those nitty-gritty theoretical investigations, what do we gain if we have little hope of applying them in the field? Well, the theoretical knowledge provides guidelines for laboratory experiments, especially for how to design and conduct them. So, let us look into this subject.

8.7.4 Application to Laboratory Studies

Usually, a laboratory study of predator–prey interaction process is described in a pair of series observed over a certain period of time in various lengths. The problem here is that, in general, the variability of the data points in an observed series would not cover an adequate section of the reproduction surface of either species that we wish to estimate, i.e. condition (ii) in the foregoing section may not be met. This is particularly true with a pair of well-regulated series about their own equilibrium levels. That is, the data points (in either species) tend to congregate only around a small section of the species' conditional curve about the equilibrium point. In other words, those data points, comparable to Figures 8.9 and 8.10 graphs, would form a blob around a central area with which we would not see the conditional curve. Two ways to get around the problem are conceivable.

One way is to directly create a conditional reproduction curve in the manner that Professor Utida did in his discrete experiment which I described with respect to Figure 6.3a in Chapter 6: this could be done by varying the density of one species while fixing the density of the other species. The other way is to expand the range of variation in data points in a series experiment by varying the initial (starting) pair of densities well away from their own equilibrium levels. An adequate number of separate, short-term series with varied initial densities would hopefully provide good information for guessing the endogenous reproduction surfaces of both species.

Now, the preceding arguments have been based entirely on the models that I developed from scratch, despite there having been quite a few mathematical models of predator–prey interaction processes. Nonetheless, I have used none of them. The reason is that these earlier models (that I am aware of) contain some problems that could hinder an analytical study. So, I should reveal them rather than ignore them.

8.8 Problems Inherent to Earlier Models

The familiar Lotka–Volterra model (published independently in 1925 and 1926) is the first and simplest form of a mathematical model of predator–prey interaction processes, given in the form of a pair of differential equations:

$$dx/dt = x(\alpha - \beta y) \quad \text{for prey X}$$
$$dy/dt = y(\delta x - \gamma) \quad \text{for predator Y}$$

in which the derivative on the left-hand side for each of X and Y is the instantaneous rate of change in the respective population whose dynamical behaviour is controlled by the two parameters (in Greek letters) on the right-hand side. The model was historically important in that, for the first time in the development of modern ecology, it showed that a predator–prey interaction could be cyclic. It elicited a serious attention of ecologists to a theoretical approach to the interaction processes. Nonetheless, some mathematicians and ecologists have been concerned about a few unrealistic aspects of this classical model.

First of all, in the (practical) absence of the predator (i.e. y is near 0), the prey population would increase at the constant rate α, which is a Malthusian process and unlikely to happen in the real world. Second, in the absence of prey (i.e. $x = 0$), the predator would survive for a while (depending on the value of γ), even though heading for an eventual extinction. Nonetheless, these are not serious faults in that the model could be interpreted as a linear approximation of the behaviour of a natural population in the very vicinity of its equilibrium level.

The real problem is that the model conceals a detrimental attribute, although not so well publicized in ecology. It has been known among applied mathematicians that the iconic oscillatory motion of this pioneering model is initial-state dependent. This means that, given an initial set of densities (x_0, y_0), the subsequent oscillatory motion is completely determined and conserved for good unless somehow altered by an external force. This attribute can be seen clearly in a phase-space representation, the plot of y on x, which exhibits a closed orbit *apparently* similar to the limit cycle in Figure 8.2a$_2$. However, the similarity is superficial in that, unlike a limit cycle, the Lotka–Volterra cycle does not converge to a fixed orbit. Rather, each time starting with a particular set of initial state, it forms its own closed orbit. Thus, the Lotka–Volterra model process generates a number of different orbits, all of which are independent but concentric to each other about the central point which, in turn, is determined on its own by the four parameters as well as by the pair of initial densities.

It sounds like an innocent property, but would in fact cause the system to become wildly uncontrolled under exogenous influences. It is similar to the neutral equilibrium state in a single-species process that exhibits a random walk under exogenous influences, as already explained in Chapter 7. In other words, the Lotka–Volterra model is fragile and produces unregulated dynamics under exogenous influences: it would not meet the stipulation equivalent to (7.6-ii) for ensuring a persistent state. So, even though an epoch-making device, the model is of little practical value as an analytical tool in modern ecological studies.

A decade later, another historically important model was published by the Irish-Australian entomologist A. J. Nicholson with the collaboration of V. A. Bailey, an applied mathematician and a physics professor at the University of Sidney (Nicholson and Bailey, 1935). It showed that a predator–prey interaction tends to be cyclic. Unlike the Lotka–Volterra model (which is in the form of differential equations and tends to hide some details of the population processes), the Nicholson–Bailey model is in a difference-equation form, incorporating the way a predator catches its prey along a search path. Unfortunately, the model (in its endogenous form) had a serious problem as already well known among ecologists: its equilibrium point is unstable, generating an oscillatory motion with ever-increasing amplitudes. The conditional reproduction curves of the model (that I introduced in the present chapter) readily reveal the causes of this anomaly: the conditional rate $r(x_t | y_t)$ of the prey is independent of density x_t, i.e. the respective conditional curve is flat across the variations in x_t. As well, the conditional rate $r'(x_t | y_t)$ of the predator linearly increases as x_t increases without bound, violating the ecological stipulation that it should have an upper limit as shown in Figure 8.7.

A few more models have come to my attention since the mid-1960s in which the authors tried to tame the wild Nicholson–Bailey model. Each work found a way to make the equilibrium level stable but at the cost of introducing another type of anomaly. Again, the conditional reproduction curves of these models reveal that all of them failed to fix the unbounded $r'(x_t | y_t)$ of the predator Y: cf. figure 4.7 of my previous book (Royama, 1992). [Note: In that figure 4.7, the notations (x, r) and (y, r') are the reversal of those in the present chapter.]

The aforementioned anomalies may not be an issue if you use these models entirely for descriptive or forecasting purposes. However, if an ecological model is used as an analytical tool, it should be built upon ecological first principles. Otherwise, it may (or is even likely to) mislead you at some point in an application. After all, as compared with the models I developed in this chapter, the earlier models are not simpler in algebraic structures and are not advantageous in a practical use. So, I suggest that these be displayed in a history museum.

So far in the present chapter, we have investigated interaction processes between single predator species and single prey species. However, this situation rarely occurs in the field: most species belong to a system of food webs and interact with many other species. Then, what would we expect to see?

8.9 Interactions between Predator Complex and Prey Complex

Long ago, I went to the University of Wisconsin to give a research talk. There, I had an opportunity to visit with Professor Lloyd B. Keith and his colleagues, well-known experts in the field of wildlife management. Through conversations with them, I learned that a certain group of predator species appeared to act like a single unit (or complex), preying on a group of prey species as another complex. As a result, these species tend to cycle in unison, much like cyclic interactions between a single-predator and a single-prey species.

220 · Predator–Prey Interaction Processes

The Wisconsin people gathered their population data from various sources over wide areas along the boreal forest zone of North America, e.g. the pelt-harvest statistics, reports from hunting-license holders, or results compiled from regional inquiries on relative abundance of certain species, etc. I summarized these findings in figures 6.22 and 6.23 (pp. 229–230) of my previous book (Royama, 1992). Although unlikely to be accurate in detail, what has been revealed is quite remarkable. In particular, it suggests that the results of the foregoing theoretical investigation of the single-predator versus single-prey interaction should apply in principle to a complex versus complex interaction process in the field. Nonetheless, I saw a noticeable difference: the phase lag in the interaction cycles between the complexes in the field was not as pronounced as between a single predator-vs-single prey interaction in the foregoing theoretical study. The cause of the difference, as I conceive, is as follows.

In the theoretical two-species interaction, the survival and reproduction of the predator depend solely on the abundance of its prey, and vice versa. Then, the predator responds to a change in the abundance of the prey primarily through the reproduction of offspring, known commonly as 'numerical response' in population ecology. This results in the predator cycle lagging behind the prey cycle. In the field, however, the predator (individually or as a species) has the freedom to switch from one prey species to another according to whichever is more (or less) profitable at any given moment, just as I observed with my titmice (cf. Chapter 1) or with the complex of natural enemies (predators and insect parasitoids) that utilize spruce budworm (*Choristoneura fumiferana*) at its various developmental stages (Royama *et al*. 2017). In other words, a predator (parasitoid) can respond to changes in the abundance of a given prey (host) species quickly (without delay) in time. Such quick responses must have resulted in a much reduced phase shift between the complex-vs-complex cycles: the shift may be so reduced that even a predator cycle precedes the corresponding prey cycle because these data tend to be subject to large sampling errors.

From the above point of view, the notion of interactions among species complexes may shed light on the well-known but not well-understood phenomenon: wildlife's 10-year cycle.

8.9.1 The Myth of Wildlife's 10-Year Cycle

As the title of Lloyd Keith's book (1963) *Wildlife's Ten-year Cycle* indicates, the 10-year cycle is a ubiquitous, well-recognized occurrence, especially among hunters and trappers. I have reviewed some early theories in section 5.3 (pp. 176–8) of my previous book (Royama 1992): many of them were no more than shear speculations, not worthy of consideration. Still, even today, many people think it rather mysterious or enigmatic: why are the 10-year cycles so ubiquitous? As I see it, nothing is enigmatic. Recognizing the interaction between the two complexes, combined with the foregoing theoretical investigations, we understand the nature of the phenomenon quite clearly.

The 10-year cycle (on average about 10, actually) is primarily the consequence of a particular combination of the parameter values, e.g. the set (r_m, r'_m, b, b') of model (8.4). The lynx cycle that I mimicked earlier in the present chapter is an example. Many other predators, mostly mammals but some raptors (not from the Jurassic Park but the real birds of prey, e.g. hawks and owls), with similar characteristics in terms of the set of parameter values (r'_m, b') would hunt the prey, comprised of hares, rabbits, and grouse(s) with a (presumably) similar set of parameter values (r_m, b). These are all results of their life histories that must have evolved under certain biological characteristics and environmental conditions that they share, although the details are largely unknown and would be a good subject of future studies. Because so many predator species are involved to form a complex, together with an apparent limit-cycle regularity (which happened to be generated by the particular set of parameter values), it has drawn the special attention of hunters and trappers. Hence the myth, I suppose.

With the above notes on the wildlife cycles, I now leave the present theme and move onto Chapter 9 on the interactions among competing species.

Appendix 8A: Ecological Mechanism Underlying the Equation $q(x_t) = [1 - \exp(-bx_t)]$ in (8.2)

Let us do a thought experiment. A number of points (as prey individuals) are distributed at random over the floor of your room, like Figure 3.1 in Chapter 3. You toss a ring (a predator individual) at random over the floor for a given (fixed) number of times per unit time. Record the number of successful trials, i.e. those tosses that successfully 'captured' at least one individual prey.

Let us assume that, for each search session (one toss of the ring), the predator can capture only one individual prey at a time, and that, having consumed that one, it makes a new search (another ring toss) elsewhere independent of the preceding search, i.e. similar to what my great tits did within a given hunting site (cf. Chapter 1). In other words, the number of successful tosses is the number of prey captured by the predator individual per unit time, or per unit effort, i.e. the fixed number of ring tosses, given the number of dots on the floor. [Note: Let us assume for simplicity that the dot is not removed but stays there on the floor as it has been.]

Assume further that the dots (prey) on the floor are Poisson-distributed with density x and the size of the ring is s, such that the expected (average) number of dots in a tossed ring is sx. Then, the probability of capturing i prey individuals at a toss is given by formula (3.6) in Chapter 3, i.e. $\Pr(i) = (sx)^i e^{-sx}/i!$. Then, the probability of capturing none (i.e. $i = 0$) is $\Pr(0) = e^{-sx}$ because $(sx)^0 = 1$ and $0! = 1$. Also, a toss of the ring will capture either 0 or at least one dot with probability 1 because nothing else occurs. So, $1 - \Pr(0) = (1 - e^{-sx})$ gives the probability of capturing at least one prey individual at a toss. Thus, the number of prey each predator individual can capture per unit time is expected to be proportional to $(1 - e^{-sx})$.

Now consider that the density x in the above trial be the density of prey at the t-th generation, i.e. x_t. Then, the foregoing expression $(1 - e^{-sx})$ is written as $1 - \exp(-sx_t)$, and replacing the ring size s with predator's hunting efficiency b', we find $q(x_t) = 1 - \exp(-b'x_t)$.

Appendix 8B: How to Find the Equilibrium Levels of the X and Y Series in the Simultaneous Equations (8.4)

The equations are repeated here for convenience:

$$r_t(x_t, y_t) = [r_m \exp(-x_t)] \exp(-by_t) \quad \text{for prey X} \quad (8.4a \text{ rpt})$$

$$r'_t(y_t, x_t) = [r'_m \exp(-y_t)][1 - \exp(-b'x_t)] \quad \text{for predator Y.}$$
$$(8.4b \text{ rpt})$$

Let x^{**} and y^{**} be the equilibria for prey X and predator Y. [Note: x^{**} or y^{**} here is distinguished from the equilibrium x^* in a single-species process in the earlier chapters.] Now, at an equilibrium state, the rate of change in population is equal to 1 such that we find from (8.4a):

$$\exp(-x^{**}) = 1/[r_m \exp(-by^{**})] \quad (8\text{B}.1\text{a})$$

Appendices · 223

$$\exp(-b'x^{**}) = [r'_m \exp(-y^{**}) - 1]/[r'_m \exp(-y^{**})]. \quad (8B.1b)$$

Because $\exp(-b'x^{**}) \equiv [\exp(-x^{**})]^{b'}$ [see Note below], substituting this in (8B.1b), we find:

$$\exp(-x^{**}) = \{[r'_m \exp(-y^{**}) - 1]/[r'_m \exp(-y^{**})]\}^{1/b'}. \quad (8B.2)$$

[Note: Your homework. Hint: ln-transform both sides to see what happens.]

Equating the right-hand side of (8B.2) to the same side in (8B.1a), and after some manipulations, we have:

$$[r'_m \exp(-y^{**})] = r_m^{b'}[\exp(-y^{**})]^{bb'}[r'_m \exp(-y^{**}) - 1]. \quad (8B.3)$$

Then, solving (8B.3) for $\exp(-y^{**})$, we would find y^{**} in terms of the parameters (r_m, r'_m) and (b, b'). The problem here is that a simple algebraic expression of the solution for y^{**} in (8B.3) does not in general exist. Fortunately, a numerical solution by computer is readily found, as shown below.

Notice that the expressions in both sides of (8B.3) are decreasing functions of y^{**}, but they decrease at different rates. So, if we plot each side of (8B.3) against y^{**} (treating it as a variable), we would see the two curves intersecting (crossing) each other at a certain point, and the reading of the point on the vertical axis gives the numeric value of exp $(-y^{**})$ and hence y^{**}. Then, the substitution of it in (8B.1) would evaluate x^{**} as well. In the Figure 8.2a example in the main text, the parameter values are: (r_m, r'_m) = (2.0, 2.7) and (b, b') = (5, 6). So we find: (x^{**}, y^{**}) = (0.170442, 0.137135). These are exact for the five digits below the decimal point.

Appendix 8C: How to Generate Correlated Series of Random Numbers

Let $\{u_i\}$ and $\{v_i\}$ be the two iid series of random numbers with mean 0 and variance σ^2 and let α and β be certain positive constants related by the constraint $\alpha^2 + \beta^2 = 1$. We now create a new series $\{w_i\}$ by the sum $\alpha u_i + \beta v_i = w_i$, which is also an iid with mean 0 and variance σ^2 for all i. [Note: in the following, the suffix i will be dropped.] The mean of w is zero because $E(w) = E(\alpha u + \beta v) = \alpha E(u) + \beta E(v) = 0$: cf. Appendix 7A in Chapter 7.

As for the variance, notice that

$$\text{Var}(w) = \text{Var}(\alpha u + \beta v)$$
$$= \alpha^2 \text{Var}(u) + 2\alpha\beta \text{Cov}(u, v) + \beta^2 \text{Var}(v).$$

However, $\text{Cov}(u, v) = 0$ because u and v are uncorrelated as assumed. Also, as assumed, $\text{Var}(u) = \text{Var}(v) = \sigma^2$. Hence, $\text{Var}(w) = (\alpha^2 + \beta^2)\sigma^2 = \sigma^2$ because $(\alpha^2 + \beta^2) = 1$ as defined.

Now the (Pearson) correlation coefficient between u and $w = (\alpha u + \beta v)$ is given by:

$$\text{Cov}(u, \alpha u + \beta v) / [\text{Var}(u)\text{Var}(\alpha u + \beta v)]^{1/2}.$$

But the numerator $\text{Cov}(u, \alpha u + \beta v) = \alpha \text{Var}(u) + \beta \text{Cov}(u, v) = \alpha\sigma^2$ because $\text{Var}(u) = \sigma^2$ and $\text{Cov}(u, v) = 0$. Also, the denominator $[\text{Var}(u)\text{Var}(\alpha u + \beta v)]^{1/2} = (\alpha^2 + \beta^2)\sigma^2 = \sigma^2$ because $(\alpha^2 + \beta^2) = 1$. Hence, the correlation coefficient between w and u is given by $\alpha\sigma^2/\sigma^2 = \alpha$. Likewise, w is correlated with v by as much as β.

9 · *Interspecific Competition Processes*

9.1 Preamble

The subject I deal with here has been considered to be one of the major ecological principles, often referred to in many standard textbooks as *the competitive exclusion principle* or Gause's principle (law), named after the Russian biologist Georgii Frantsevich Gause, who conducted well-known laboratory experiments (on competition among some ciliates) and the author of *The Struggle for Existence* (Gause, 1934). The principle can be stated (or taught in an ecology class) as: *No two species of similar ecology can coexist or occupy the same ecological niche.* Some observations (in the field or in the laboratory) that followed Gause have more or less agreed with him, while others have shown or argued otherwise. On the whole, in the current ecology, there seems to be considerable skepticism about the 'principle'.

My aim here is to theoretically study the nature of interaction processes between two competing species, hoping to find general principles behind the processes. I have no previous publications on this subject during my active career, but I have kept it in mind for some time now. It is a natural extension of the preceding chapter on the two-species (predator–prey) interaction processes. So, I decided to include this subject in the present book. Let me begin with formulating an appropriate model.

9.2 Formulation of Competition Model

The model I consider here is identical in structure to the predator–prey process model (8.1), in that both are two-species interaction processes. I repeat (8.1) here for your convenience:

$$r_t(x_t, y_t) = [r_m \exp(-cx_t)] p(y_t) \qquad \text{(8.1a rpt)}$$

$$r'_t(y_t, x_t) = [r'_m \exp(-c'y_t)] q(x_t) \qquad \text{(8.1b rpt)}$$

in which x_t and y_t are the population densities at time step t (e.g. generation) of the prey species X and the predator species Y, respectively. The ecological meanings of the parameters r_m, r'_m, c, and c' are exactly the same as those originally developed for the single-species process in Chapter 3. In particular, (r_m, r'_m) are the potential reproductive rates of the species X and Y, realizable when no fellow (within-population) competitors are present; (c, c') are measures of the intensity of competition within each specific population.

The competition model here differs from the predator–prey model (8.1) with respect to the ecological meanings of the functions $p(y_t)$ and $q(x_t)$. In the predator–prey model, $p(y_t)$ was the adverse effect of predation on the reproductive rate of the prey species and was a decreasing function of y_t, whereas $q(x_t)$ was the beneficial effect of prey abundance on the rate of predation and was an increasing function of x_t. In the present theme of competition, each function characterizes the adverse effect of the presence of one species on the reproductive rate of the other species. In other words, $q(x_t)$ should be a decreasing functions of x_t inasmuch as $p(y_t)$ is a decreasing function of y_t. So, what is an appropriate form for each function?

I insist that they should be based on ecological principles but algebraically simple, such that they should satisfy the following two attributes. First, as in the predator–prey interaction, both $p(y_t)$ and $q(x_t)$ must be defined in the positive domain no greater than 1 because neither the reproductive rate $r_t(x_t, y_t)$ nor $r'_t(y_t, x_t)$ should exceed r_m or r'_m. Second, $p(y_t)$ and $q(x_t)$ should be monotonically decreasing functions of y_t and x_t, respectively, and asymptotically tend to 0 as y_t and x_t increase indefinitely. I find that the exponential function of the form 'exp($-u$)', in which the generic variable u stands for population density, would satisfy the above stipulations. As well, there is no reason to consider that the two functions should be of different forms. Thus, I suggest:

$$p(y_t) = \exp(-by_t)$$
$$q(x_t) = \exp(-b'x_t),$$

in which b (or b') now is a measure of the *sensitivity* of the species X (or Y) to the presence (adverse effect) of the species Y (or X): in particular, the higher the value, the more sensitive. [Note: The above formulae assume the same form as $p(y_t)$ in (8.3) for the predator–prey interaction. Although the two species would not devour each other in the competitive interaction, I use this particular algebraic form as a simple device to describe the adverse effect (on the reproductive rate of either X or Y) in either category of interactions.]

Substituting the above formulae of p and q for those in (8.1 rpt), and after a little algebraic rearrangement, the full formulae for the two competing species are given by:

$$r_t(x_t, y_t) = [r_m \exp(-by_t)] \exp(-cx_t) \quad \text{for species X} \quad (9.1a)$$

$$r'_t(y_t, x_t) = [r'_m \exp(-b'x_t)] \exp(-c'y_t) \quad \text{for species Y.} \quad (9.1b)$$

[Note: Recall that, in Chapter 8, the parameters c and c' in model (8.1) were set equal to 1 to reduce the number of parameters. In the present theme, I retain these parameters explicitly as they play important roles.]

Now, recalling that $r_t = x_{t+1}/x_t$ and $r'_t = y_{t+1}/y_t$ (as defined in Chapter 8), let us try a few simulations to see how the model processes (9.1) would behave.

9.3 Simulations

Where should we start? To decide, let us take a look at a result of the classical Gause's experiment (figure 22 in Gause, 1934) with two Paramecia species, *Paramecium aurelia* and *P. caudatum*, which were cultured together in the same medium with the bacterium (*Bacillus pyocyaneus*) as common (shared) food. In Figure 9.1, I reproduced Gause's original graphs, copying them by hand as accurately as I could. [Note: The data points and the shape of the curves are as accurate as I could trace. However, I rearranged the original graphs in two different formats: (a) for 'Mixed populations' and (b) for 'Cultured separately'.] We see that, at the start, both species grew in a similar manner, although *P. caudatum* appeared to grow more rapidly initially than *P. aurelia*. However, *P. caudatum* somehow reached a peak at around day 8 and thereafter went down with no apparent sign of recovery. In the meantime, *P. aurelia* continued growing to have apparently reached a plateau. By repeating the experiments with variations in the culture media, Gause demonstrated that *P. caudatum* was always eliminated.

To comprehend the patterns of Gause's experiments in terms of model (9.1), I did some simulations with different sets of the parameter values. Figure 9.2 shows one of those trials that mimics the observed patterns quite well. This encouraged me to do some more trials, although not quite systematically, to find that every trial invariably ended up with the elimination of one species. So I thought my model (9.1) did confirm *the competitive exclusion principle* until I tried a few more simulations to make sure. Then, I found that, in certain combinations of the parameter values,

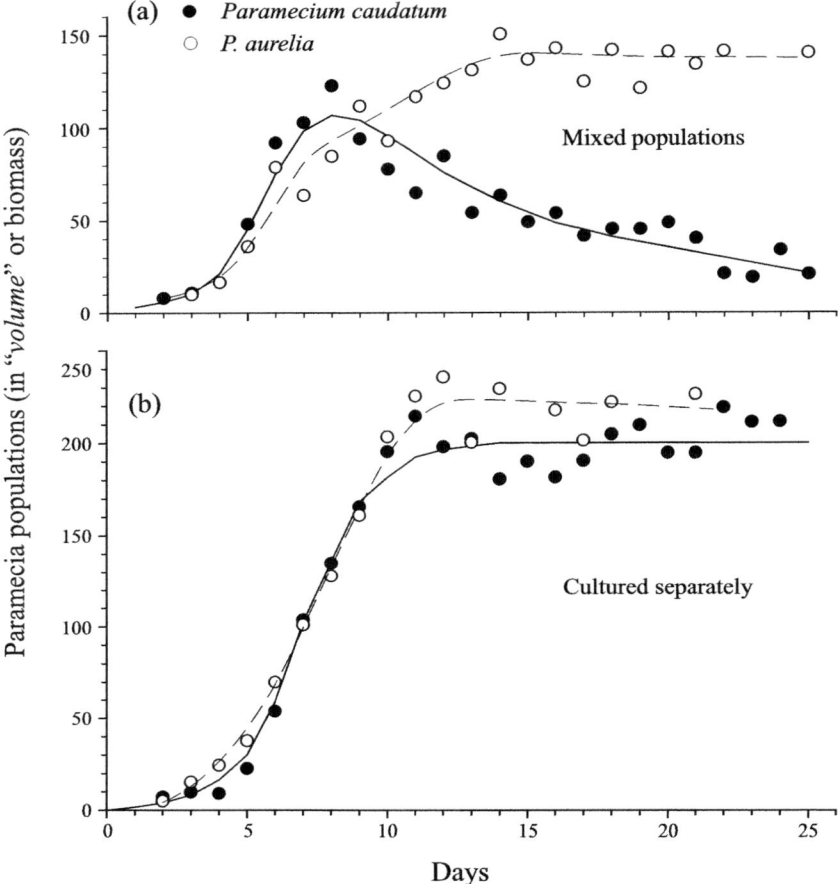

Figure 9.1 Gause's famous experiment with the two Paramecia species, *Paramecium aurelia* and *P. caudatum*. Hand-copies of figure in Gause (1934)

the two species did coexist, as exemplified in Figure 9.3. This motivated me to systematically find the criteria for how and when the two species could coexist or, conversely, one species would be eliminated.

9.4 Criteria for Coexistence and Elimination

9.4.1 Algebraic Properties of Criteria

To this end, let us take a look at the algebraic properties of the model processes (9.1). In particular, let us find equilibria of the series $\{x_t\}$ and

9.4 Criteria for Coexistence and Elimination · 229

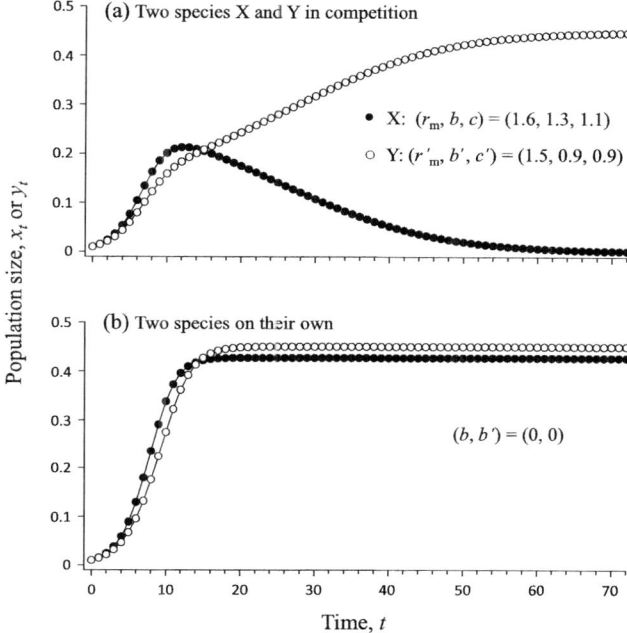

Figure 9.2 The results of a simulation with model (9.1) that mimics the Gause's experiment shown in Figure 9.1.

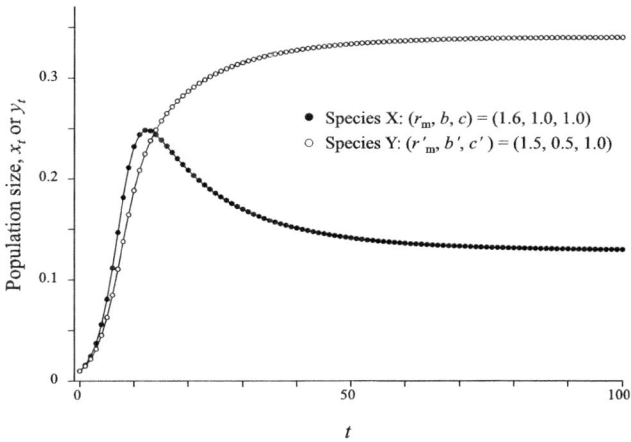

Figure 9.3 An example of simulation with model (9.1) that shows the coexistence of two competing species.

$\{y_t\}$. This can be readily done if we transform (9.1) into natural logarithms. So, writing $\ln(r) = R$ as before, we have:

$$R_t = R_m - by_t - cx_t \tag{9.2a}$$

$$R'_t = R'_m - b'x_t - c'y_t. \tag{9.2b}$$

[For the sake of algebraic convenience, I leave the variables x_t and y_t unconverted to logarithms.] At their equilibrium densities, designated x^{**} and y^{**}, it must be that $R_t = R'_t = 0$, i.e. no more population changes. Thus, we have:

$$0 = R_m - by^{**} - cx^{**} \tag{9.3a}$$

$$0 = R'_m - b'x^{**} - c'y^{**}. \tag{9.3b}$$

Solving the above for x^{**} and y^{**} after a little algebra shown in Appendix 9A, we find:

$$x^{**} = (c'R_m - bR'_m)/(cc' - bb') \tag{9.4a}$$

$$y^{**} = (cR'_m - b'R_m)/(cc' - bb'). \tag{9.4b}$$

[Note: As already mentioned in Appendix 8B in Chapter 8, the notation x^{**} (or y^{**}) is distinguished from its counterpart x^* in a single-species process in the earlier chapters.]

I now show that relationships (9.4) provide information on the algebraic prerequisites (stipulations) for the coexistence of X and Y or the elimination of either species. In particular, we look into the quotients on the right-hand sides of (9.4) to notice that the denominator (designated D) common to x^{**} and y^{**}, and each of the numerators (N_x for x^{**} and N_y for y^{**}) can be positive or negative, excluding $D = 0$ for now. Table 9.1 lists all possible combinations and arrangements of these algebraic components being positive (+) or negative (−) in five major categories (with two subcategories a and b in the last three).

The above attributes of x^{**} and y^{**} in terms of their signs are basically as follows. The positive sign indicates the existence of the equilibrium. In particular, in categories (i) and (ii), both x^{**} and y^{**} are positive, indicating the coexistence of X and Y, if not always as we will see. The negative sign indicates that the equilibrium is not realizable with real populations, as they are non-negative entities. Thus, in categories (iii) and (iv), the elimination of one species is inevitable. Category (v) is infeasible; it cannot be physically realizable, as shown in Appendix 9B,

Table 9.1 Possible combinations of $D = (cc' - bb')$, $N_x = (c'R_m - bR'_m)$ and $N_y = (cR'_m - b'R_m)$ as positive or negative in five categories (i) to (v).

	(i)	(ii)	(iiia)	(iiib)	(iva)	(ivb)	(va)	(vb)
D	+	−	+	+	−	−	+	−
N_x	+	−	+	−	+	−	−	+
N_y	+	−	−	+	−	+	−	+
x^{**}	+	+	+	−	−	+	−	−
y^{**}	+	+	−	+	+	−	−	−

and hence need not be considered. Now, let us look into each of the four feasible categories in detail.

9.4.2 Details of the Categories in Table 9.1

Category (i). The stipulations that characterize this category are: $D = cc' - bb' > 0$; $N_x = (c'R_m - bR'_m) > 0$; and $N_y = (cR'_m - b'R_m) > 0$. After a little manipulation of the latter two stipulations, we find:

$$b < c'(R_m/R'_m) \qquad (9.5a)$$

$$b' < c(R'_m/R_m). \qquad (9.5b)$$

Substituting relationships (9.5) in (9.4), we find both x^{**} and y^{**} are positive in this category. It follows that stipulations (9.5) ensure the coexistence of X and Y.

Let us do some simulations to see how X and Y would behave by varying their parameter values but within the range of variation that complies with stipulations (9.5). To begin, notice that the parameter sets (R_m, c) and (R'_m, c') on the right-hand sides of (9.5) characterize the processes of X and Y when they are on their own, i.e. not in a competitive situation. So, let's arbitrarily fix their values at: $R_m = \ln(1.6)$, $R'_m = \ln(1.5)$, $c = 1.1$, and $c' = 0.9$. I chose these values conveniently for illustration. We now vary the values of the parameters (b, b') within the ranges that comply with (9.5) to see what happens.

Figure 9.4 shows some examples, each starting with an arbitrary set of initial densities (x_0, y_0). We see that $b > b'$ results in $x^{**} < y^{**}$ in Figure 9.4a, whereas $b < b'$ results in $x^{**} > y^{**}$ in Figure 9.4b. It looks as though whichever is larger (or smaller) between x^{**} and y^{**} depends on

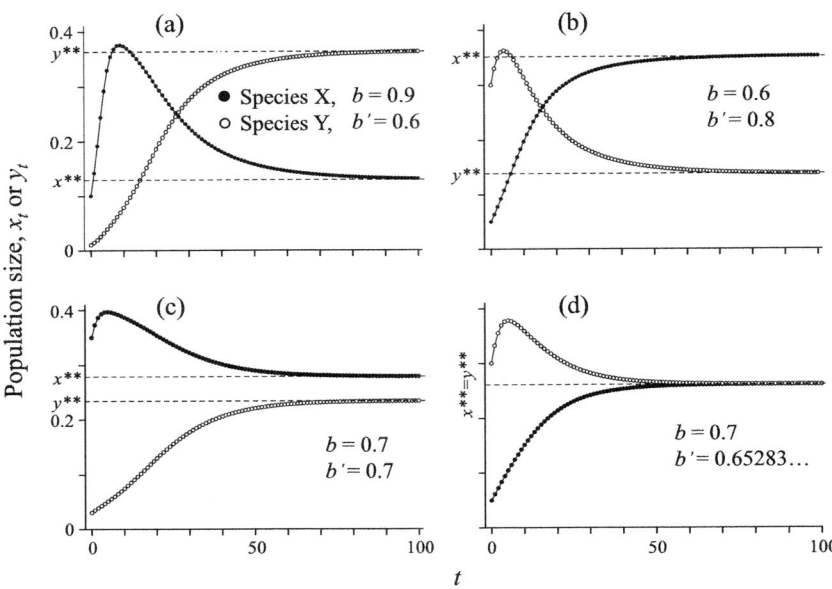

Figure 9.4 Varieties of dynamical pattern that model (9.1) generates in category (i) of Table 9.1, rendering the coexistence of the competing species X and Y.

whichever is smaller (or larger) between b and b'. However, in Figure 9.4c, $x^{**} > y^{**}$, even though $b = b'$. So, there must be a particular pair of values (b, b') at which $x^{**} = y^{**}$. This can be found in the following way. Notice that the equality $x^{**} = y^{**}$ means $N_x = N_y$ or $(c'R_m - bR'_m) = (cR'_m - b'R_m)$, and solving this equality for b', we find:

$$b' = (b+c)(R'_m/R_m) - c'. \tag{9.6}$$

Figure 9.4d shows the result of a simulation in which $b = 0.7$ and $b' = 0.65283...$ to comply with relationship (9.6), which results in $x^{**} = y^{**} = 0.26111...$ Altogether, we see the rule that, for $b' < =$ or $> [(b+c)(R'_m/R_m) - c']$, we have $x^{**} > =$ or $< y^{**}$, respectively.

The most significant attribute of category (i) is that, no matter where the X and Y populations have started, they end up converging to their respective equilibrium densities x^{**} and y^{**}, both being positive. Thus, x^{**} and y^{**} here are stable equilibria realizable in a competitive situation, and the stability ensures the coexistence of X and Y under the influence of random exogenous influences (fluctuating environment) in natural situations. [Note: The equilibrium density x^{**} (or y^{**}) can be

9.4 Criteria for Coexistence and Elimination · 233

close to 0. Then, the accidental extinction of X (or Y) could happen if the variation in exogenous influences was large enough.]

But, after all, what do stipulations (9.5) imply ecologically? At this stage of study, I am not ready to offer a precise interpretation of the stipulations. However, as a special case, consider that the two species X and Y have *similar* ecology (when on their own), i.e. $R_m \approx R'_m$ and $c \approx c'$. Then, (9.5) implies that the two species would likely coexist if each one tends to be more tolerant of the presence of its opponent species than of its own fellow individuals: that is, if interspecific competition is less intense than intraspecific competition.

Category (ii). Here, both x^{**} and y^{**} are positive as in category (i). However, unlike (i), all of the algebraic components (i.e. D, N_x and N_y in Table 9.1) are negative, i.e. $D = cc' - bb' < 0$; $N_x = (c'R_m - bR'_m) < 0$; and $N_y = (cR'_m - b'R_m) < 0$. Then, in contrast to stipulations (9.5) for category (i), we find:

$$b > c'(R_m/R'_m) \qquad (9.7a)$$

$$b' > c(R'_m/R_m). \qquad (9.7b)$$

In this situation, the equilibria x^{**} and y^{**} (albeit feasible mathematically) become unstable such that, unless the X and Y populations start exactly at $x_0 = x^{**}$ and $y_0 = y^{**}$, they diverge from the respective equilibria if displaced even very slightly, as illustrated in Figure 9.5.

In all graphs, the parameter values are conveniently set at $(R_m, b, c) = \{\ln(1.6), 2.7, 1.0\}$ and $(R'_m, b', c') = \{\ln(1.5), 1.7, 1.0\}$. Only the initial densities x_0 and y_0 are varied because the way each series diverges from the equilibrium depends solely on x_0 (or y_0) relative to the level x^{**} (or y^{**}). Let us examine each graph in detail.

In Figure 9.5a, $x_0 = x^{**}$ and $y_0 = y^{**}$ exactly, and X and Y stay at their own initial density for good if undisturbed. In Figure 9.5b, y_0 is placed slightly above y^{**} while x_0 is exactly at x^{**}. Even though y_0 is off y^{**} by as little as 0.00001 initially, the Y series diverges after a while upward away from y^{**}. As Y so diverges, X diverges downwards from x^{**}, heading for 0 to be eliminated eventually. Meanwhile, the winner Y heads for y^*, which is its equilibrium density in the absence of X. In Figure 9.5c, while x_0 is placed at x^{**} as before, y_0 is displaced slightly below y^{**}, a situation which ends up with Y being eliminated eventually. Meanwhile, as Y is disappearing, the winner X heads for x^*, i.e. its equilibrium density when X is on its own. [Note: How long each species would stay in its apparent (temporary) equilibrium state depends on how

234 · Interspecific Competition Processes

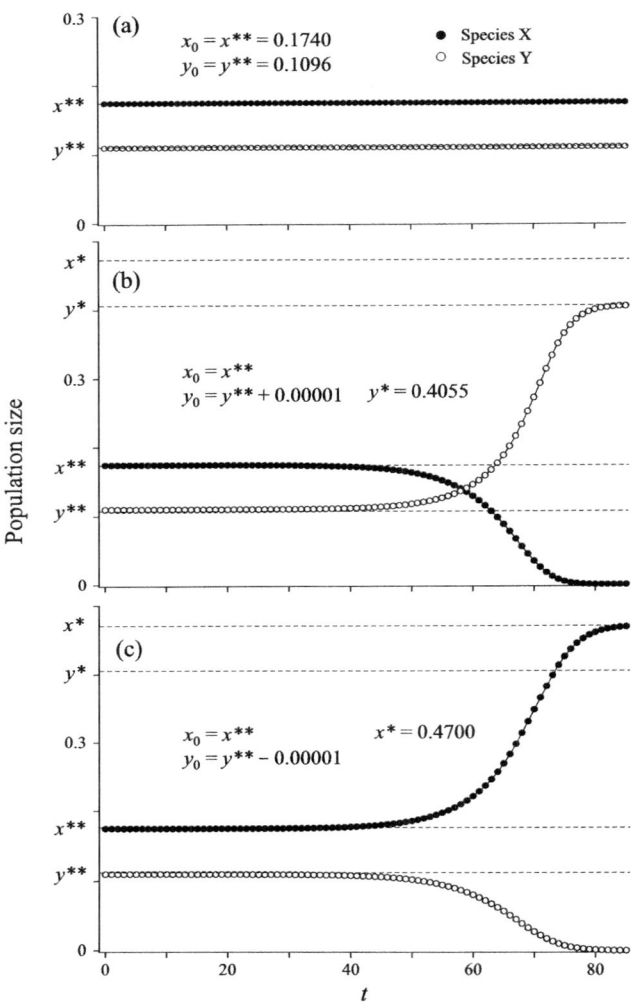

Figure 9.5 Similar to Figure 9.4 but in category (ii) of Table 9.1.

closely the initial values of both species (x_0, y_0) are to (x^{**}, y^{**}): the closer the longer, of course.] But, which species wins or loses in general situations? To find the answer, look at some more simulations in Figure 9.6, which use the same parameter values as those in Figure 9.5 but the initial values (x_0, y_0) are varied.

In the three graphs on the left, both X and Y start at low densities well below x^{**} and y^{**}. In all these graphs, y_0 is fixed at 0.01, while x_0 is

9.4 Criteria for Coexistence and Elimination · 235

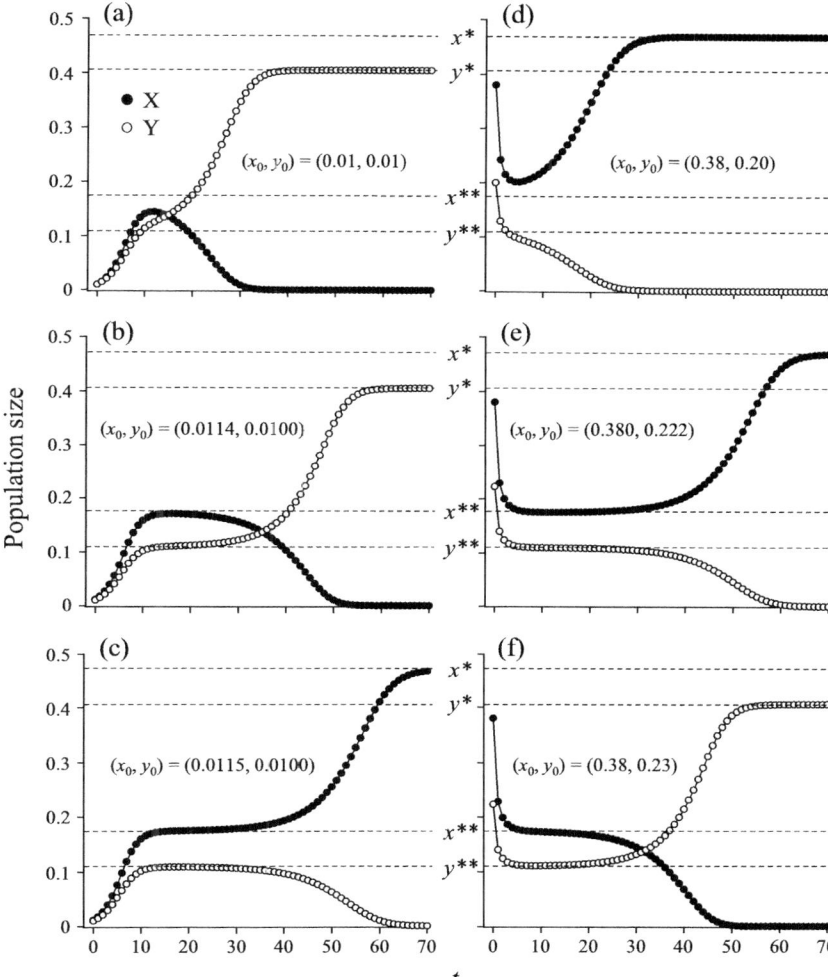

Figure 9.6 More varieties from category (ii) but the initial values (x_0, y_0) are varied.

varied as marked in each graph. Even though the variations (in x_0) are extremely small, the results are unexpectedly different. We see that Y is a decisive winner in (a). In (b), both X and Y approached x^{**} and y^{**} at more or less the same time and stay there closely for a while, but Y wins after all. In (c), both X and Y, after having approached x^{**} and y^{**}, stay there for a while as in (b), but X beats Y after all.

236 · Interspecific Competition Processes

In the graphs on the right, both X and Y start above x^{**} and y^{**}. In all graphs, x_0 is conveniently fixed at 0.38, while y_0 is varied. Results: X is a clear winner in (d); indecisive for a while in (e) but X wins in the end; and in (f), indecisive again for a while, but Y wins after all. Altogether, the simulations reveal the general tendencies: the series that locates itself above (or below) its own equilibrium level (x^{**} or y^{**}) ahead of the other will win (or lose).

Now that these simulated processes are endogenous processes, their state at a given time t, i.e. (x_t, y_t), depend entirely (deterministically) on their initial state (x_0, y_0). In other words, the fate of X or Y was already determined at the start. This does not happen in a natural environment (i.e. under exogenous influences): the fate of X or Y depends precariously on how the two series happen to behave relative to each other in the neighbourhood of x^{**} or y^{**}. That is, which species wins or loses would largely be at the mercy of Mother Nature, i.e. unpredictable. I will discuss this problem in more detail shortly.

Category (iii). This category is characterized by the algebraic attributes: $D > 0$, but N_x and N_y have different signs, i.e. if $N_x > 0$, then $N_y < 0$ and vice versa. For this reason, the category is subdivided into subcategories (iiia) and (iiib) according to the following difference in stipulations:

$$\text{(iiia)}: b < c'(R_m/R'_m) \text{ and } b' > c(R'_m/R_m), \text{ i.e., } x^{**} > 0 \text{ and } y^{**} < 0 \quad (9.8a)$$

$$\text{(iiib)}: b > c'(R_m/R'_m) \text{ and } b' < c(R'_m/R_m), \text{ i.e., } x^{**} < 0 \text{ and } y^{**} > 0. \quad (9.8b)$$

As already mentioned, a negative equilibrium means that it would not be realizable in the real world. So, what would we expect to see actually when the processes concerned comply with stipulations (9.8)?

To find out, look at the simulations in Figure 9.7, in which all parameters, but (b, b'), are fixed at $(r_m, c) = (1.6, 1.1)$ and $(r'_m, c') = (1.5, 0.9)$ as in Figure 9.4 for category (i), such that $c'(R_m/R'_m) > c(R'_m/R_m)$; and the initial state $(x_0, y_0) = (0.01, 0.01)$. Parameters (b, b') are varied as: $b > b'$ in graph (a); $b = b'$ in graph (b); and $b < b'$ in graphs (c). All of these graphs belong to category (iiia), complying with stipulation (9.8a). The last graph (d) represents (iiib), i.e. compliance with (9.8b). [Note: In category (iiib), $b > b'$ is the only possible way because $b > c'(R_m/R'_m) > c(R'_m/R_m) > b'$ to comply with stipulation (9.8b).]

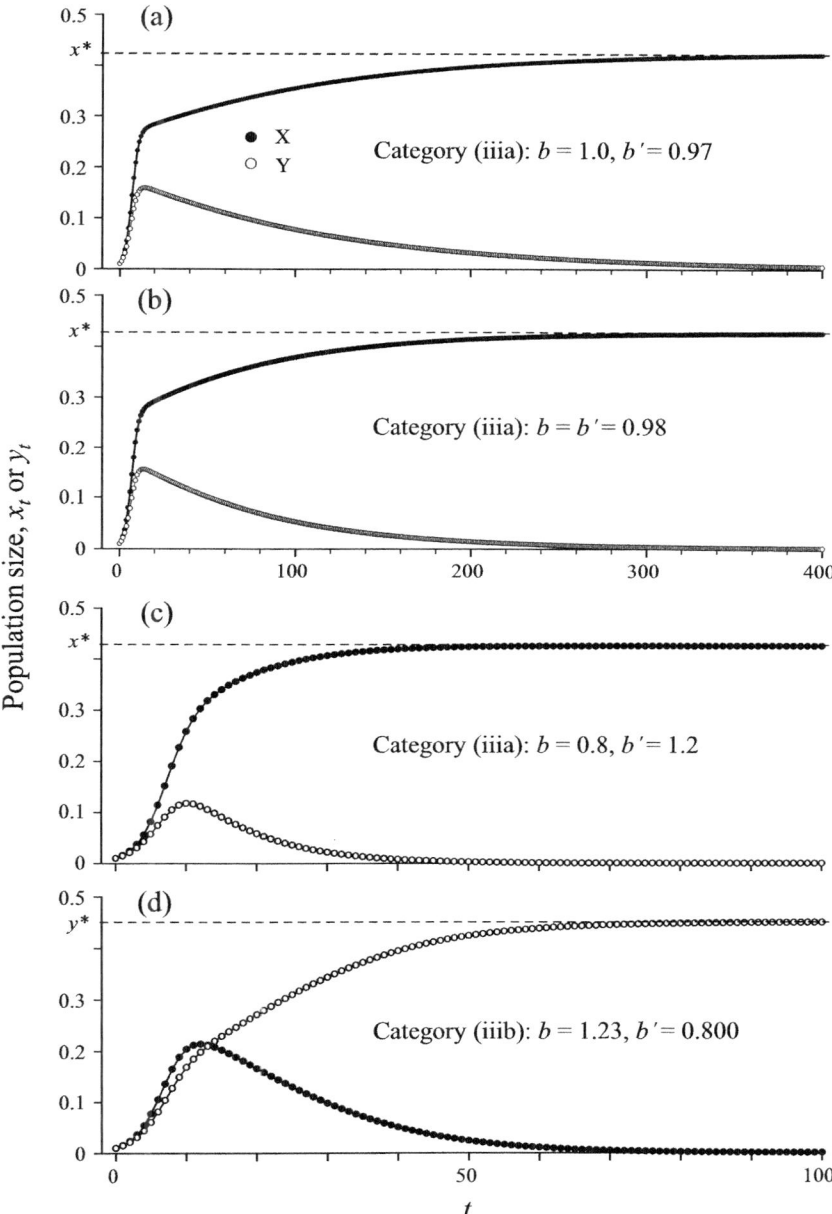

Figure 9.7 Variations in dynamical pattern generated in category (iii) of Table 9.1.

238 · Interspecific Competition Processes

Now that every possible relationship (in relative value) between b and b' is represented in Figure 9.7, it reveals the rule that the species with a negative equilibrium density (x^{**} or y^{**} in Table 9.1) would be eliminated. That is, Y is always eliminated in (iiia), whereas X is the loser in (iiib). As the loser is being eliminated, the winner heads for its equilibrium level (x^* or y^*) which is attained when there is no competing species. [Note: The length of the horizontal axis in each graph is adjusted to how long it takes for the losing species to reach the 0 level: 400 time steps in the top two and 100 in the bottom two.]

Category (iv). This category is similar to (iii) but $D < 0$ and, accordingly, the stipulations are changed to:

$$\text{(iva)}: b < c'(R_m/R'_m) \text{ and } b' > c(R'_m/R_m), \text{ i.e. } x^{**} < 0 \text{ and } y^{**} > 0 \tag{9.9a}$$

$$\text{(ivb)}: b > c'(R_m/R'_m) \text{ and } b' < c(R'_m/R_m), \text{ i.e. } x^{**} > 0 \text{ and } y^{**} < 0. \tag{9.9b}$$

Here, the rule is the reversal of that in category (iii): species with a negative equilibrium would win the competition, e.g. X wins in the (iva) situation but loses in (ivb). [Note: I do not attempt simulations for this category: you can readily confirm the tendency. Good exercises for you!] But, why do the species with a negative equilibrium win? It sounds strange because a negative equilibrium is unrealizable. How this occurs can be understood in the following way.

The species X (or Y) with $x^{**} < 0$ (or $y^{**} < 0$) in this category may, initially, increase faster than its competitor with a positive equilibrium. As this happens, the competitor would begin to lose, and the winner (with the negative equilibrium) would head directly for x^* (or y^*) which is positive. In other words, the winner heads for the real equilibrium level (with the single *), disregarding (as it were) the negative equilibrium (with the double **) as it is unrealizable.

Incidentally, the simulation in Figure 9.2, that mimics Gause's classical experiment, belongs to the category (ivb) in terms of the parameter values I used. However, it does not mean that Gause's own experimental process was in fact in that category. This is because a similar dynamical (time-dependent) pattern may occur in a different category, e.g. Figure 9.7d or Figure 9.6a.

So much for the major criteria for coexistence and competitive elimination. However, I have one more category to investigate with respect

9.4 Criteria for Coexistence and Elimination · 239

to the following attribute: although it has not been shown, among those categories in Table 9.1, there is a particular (singular) point where $D = 0$ and, in its neighbourhood, a competitive interaction process would behave in a peculiar manner.

9.4.3 Behaviour of an Interaction Process in Neighbourhood of the Singular Point

So far, we have considered the situations in which D, the denominator, common to the quotients in (9.4a and b), is either positive or negative. For your convenience, I repeat (9.4) below:

$$x^{**} = (c'R_m - bR'_m)/(cc' - bb') \equiv N_x/D \qquad (9.4\text{a rpt})$$

$$y^{**} = (cR'_m - b'R_m)/(cc' - bb') \equiv N_y/D. \qquad (9.4\text{b rpt})$$

Here, I consider what we would expect to see if $D = 0$. [Note: In reality, the likelihood of D being exactly equal to 0 is nil. However, it can be very close. Then, the behaviour of the interaction processes may be very similar to the following.]

Suppose that numerators of the quotients, i.e. N_x and N_y, are non-zero and finite-valued. Then, in the neighbourhood of $D = 0$, the quotients $N_x/D = x^{**}$ and $N_y/D = y^{**}$ become infinitely large in absolute values, a situation which carries no ecological meaning. But what if $N_x = N_y = 0$, too? We find $N/D = 0/0$, the value of which is indeterminate. However, as it turns out, we cannot dismiss this situation as ecologically meaningless, as shown below.

Because $N_x = (c'R_m - bR'_m) = 0$ and $N_y = (cR'_m - b'R_m) = 0$, solving these for b and b', we find:

$$b = c'(R_m/R'_m) \qquad (9.10\text{a})$$

$$b' = c(R'_m/R_m). \qquad (9.10\text{b})$$

Now that all of the parameters on the right-hand sides are species-specific and finite-valued, the numeric values of b and b' are unambiguously determined. However, because the quotients N/D are $0/0$, we cannot determine x^{**} and y^{**} in (9.4 rpt). In other words, we are unable to know analytically (i.e. in an algebraic manner) how the interaction process would behave.

Nonetheless, as it turns out, numerical solutions for x^{**} and y^{**} do exist: I found this fact accidentally when I did some simulations, using a

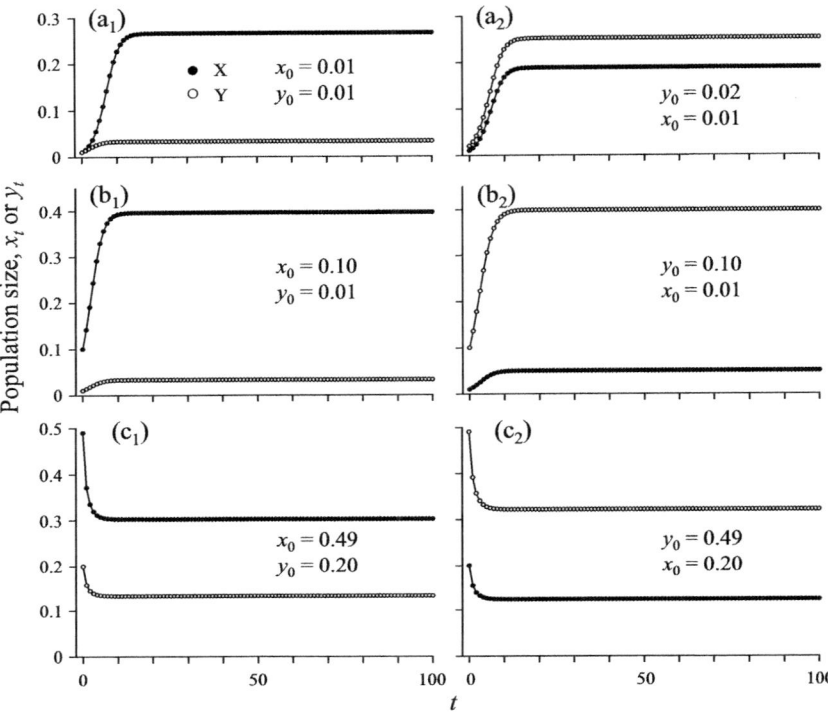

Figure 9.8 Numerical determinations of the equilibria x^{**} and y^{**} in (9.4), given a set of initial values (x_0, y_0), at the singular point where the $(D, N_x, N_y) = (0, 0, 0)$ in Table 9.1.

set of parameter values that comply with (9.10). The parameter values that I happened to use in the simulations were: $(r_m, c) = (1.6, 1.1)$ and $(r'_m, c') = (1.5, 0.9)$; such that $R_m = \ln(1.6)$, $R'_m = \ln(1.5)$; and hence from the relationships (9.10), $b = 1.04325\ldots$ and $b' = 0.94895\ldots$ Figure 9.8 shows some typical results.

Figure 9.8a$_1$ is the result of my first trial in which I started X and Y with their initial values at $x_0 = y_0 = 0.01$, expecting that my computer would say 'Computation impossible' because the equilibria given in (9.4) are 0/0, i.e. indeterminate. To my surprise (or more like astonishment), I saw the computer output a concrete picture of the series X and Y converging to (seemingly) their own equilibria x^{**} and y^{**}, at which the two species appeared to be coexisting.

For the following few days, I struggled to comprehend how this could happen. Then, a thought struck: the point at which $N/D = 0/0$ occurs as a transitional state from categories (i) to (ii), i.e. from a stable state of coexistence to an unstable state. At this transitional point, the equilibria x^{**} and y^{**} would be realizable (i.e. numerical solutions exist) when the initial values (x_0, y_0) are given: this was why the computer output the numerical results. In other words, these apparent equilibria are 'initial value-dependent' similar to the classical Lotka–Volterra model as mentioned in Chapter 8. So, I varied the initial values to confirm my interpretation. In the Figure 9.8a$_2$ trial, I changed (x_0, y_0) to (0.01, 0.02), while keeping the parameter values unchanged, which resulted in changes in the equilibrium levels. Results of a few more trials are also shown in Figure 9.8. We see that the apparent equilibria are indeed initial value-dependent.

The important issue to recognize here is that these (initial value-dependent) equilibria are fragile against exogenous influences: they exist only in their endogenous forms. In other words, in a natural environment, the apparent coexistence of the two species as we see in Figure 9.8 would be unrealizable. However, what we would actually see provides much insight, as I show now.

9.4.4 Effect of Exogenous Influences on the Transitional State of Equilibria

Let us perform some more simulations, incorporating the effect of the influences into models (9.2). To begin, using the method described in Appendix 8C, I generate a pair of (50%) correlated random series (iid, normal, mean 0 and variance 0.01), and incorporate them into the Figure 9.8a$_2$ example; Appendix 9C shows how to do this. The results are given in Figure 9.9.

Graph (a) shows the two (X and Y) series for the first 100 time steps (generations). We see that X and Y appear to still be in equilibrium, i.e. apparently coexisting. However, a further extension to 1000 time steps in graph (b) shows the eventual elimination of Y. [Note: It happened to be Y in this example. In general, which species is to be eliminated depends unpredictably on the random variation in the exogenous influences.] As competitor Y is being eliminated, X heads for the stable equilibrium level ($x^* \approx 0.43$) which is realizable when X is on its own. [Note: $x^* = (1/c)\ln(r_m)$ as shown in Section 4.3 in Chapter 4.] The above results have a certain pragmatic implication. However, before getting to it, one point should be clarified.

Statistically speaking, the probability of the set of the parameters (b, b') to take the exact values in (9.10) would be nil. However, the probability

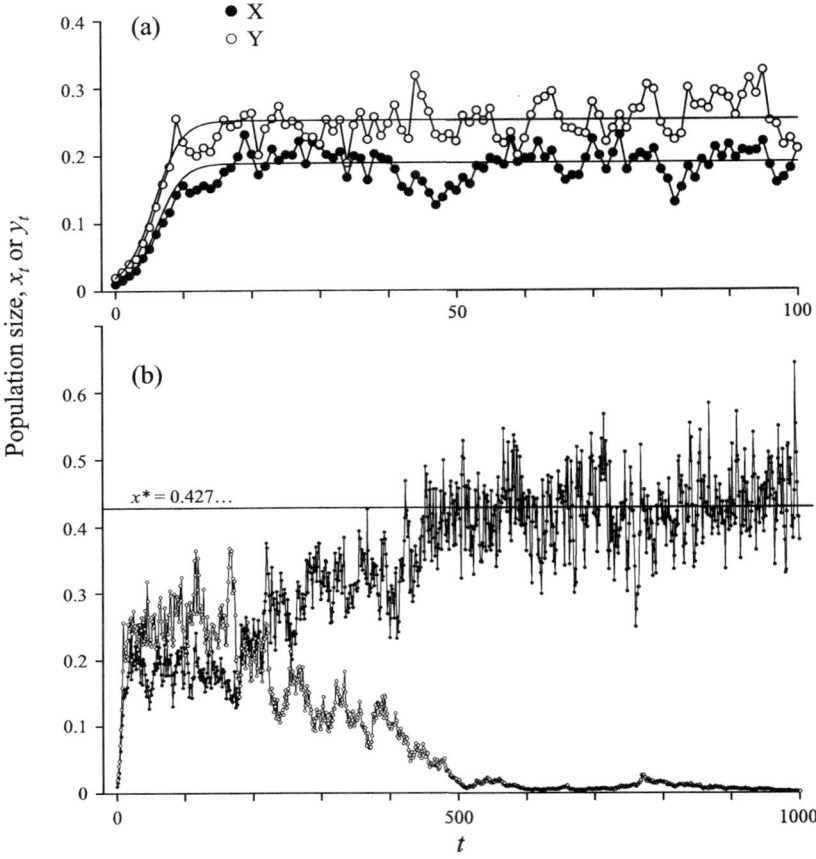

Figure 9.9 Effect of exogenous influences on the fragility of the transitional state of equilibria.

of the (b, b') values to be realized in a close proximity of (9.10) would exist. So, we should investigate the behaviour of an interaction process in the neighbourhood of the transitional point. In particular, the point of interest here is the process behaviour just inside the category (ii) in Table 9.1 in which the equilibrium densities have been shown to be unstable even in the endogenous form.

Figure 9.10 shows three examples in which the (b, b') values are offset by 0.01, 0.05 and 0.1 from their own respective (exact) values in (9.10), such that they comply with stipulations (9.7) and hence are inside

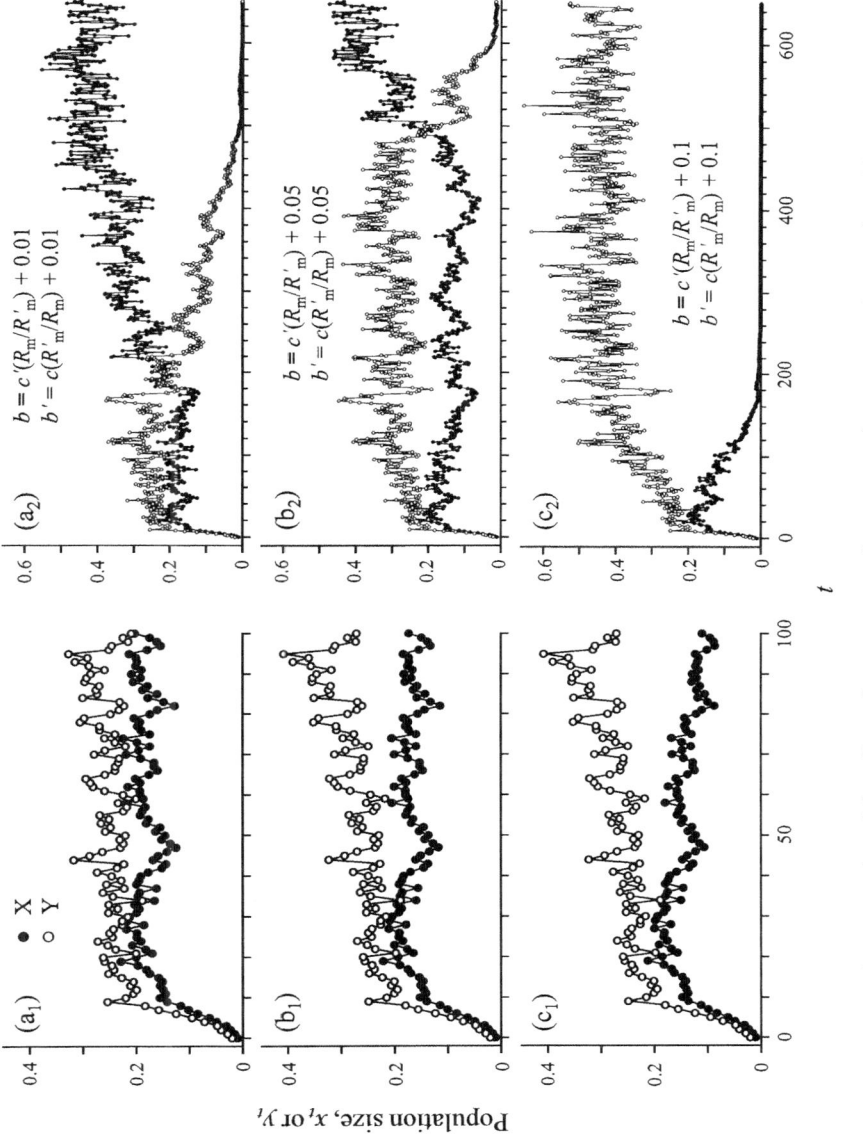

Figure 9.10 Further examples of the effect of exogenous influences on the transitional state of equilibria.

category (ii). The graphs on the left-hand side show what happened for the first 100 time steps, and those on the right show what happened after extending the simulations further to 650 steps. We see that, in all cases, one species is eventually eliminated as expected of the processes in category (ii).

However, there is one aspect in Figure 9.10 that I had not quite anticipated: it could take a long time for one species to actually be eliminated. In fact, it took 200 time steps (generations) for simulation (a_2), and as many as 500 steps for simulation (b_2), to show the first sign of elimination.

This could be a serious problem if we encountered a similar situation in an experiment. Because we would not know the (endogenous) mechanisms that had generated the observed series, we could not be sure if the two species were coexisting, or if one of them was going to be eliminated and, if so, which one and when.

The seriousness of this problem of uncertainty depends on the likelihood of the above situations occurring in the real world. I had thought that the likelihood should be low because the probability of a process to occur at or near the singular point was low. However, a situation very similar to the Figure 9.10 simulations was actually observed in the series of experiments with the two species of flour beetles, *Tribolium confusum* and *T. castaneum*, conducted by Thomas Park (1948), and analysed by P. H. Leslie, T. Park and D. B. Mertz (1968). Of many series of separate experiments, all but one series ended with the elimination of one or the other species. In the exceptional series, the two species coexisted for as long as 30 generations over 2.6 years, and were still coexisting when the experiment was terminated.

9.4.5 Uncertainty in Experimental Ecology

The similarity in results between these experiments and simulations exposes the problem of uncertainty: we would never know whether the observed (apparent) coexistence was just temporary or in a persistent state. This is a serious problem, or more a dilemma, because an experiment is carried out only over a limited (usually short) period of time, and its result may be inconclusive. If no elimination occurred, it would be impossible to determine whether the two species coexisted temporarily as in category (ii) or in a stable manner as in category (i). In other words, by experimentation, one can only confirm the elimination of one species but not coexistence.

From the above uncertainty point of view, let us also take a look at Gause's experiment on the Paramecia, reproduced in Figure 9.1. Two things are apparent. First, the *Paramecium caudatum* population has not yet been eliminated when the experiment was terminated after 25 days. Second, the *P. aurelia* population has apparently reached an equilibrium level while the *P. caudatum* population is still quite high, although it appears to be continually declining. So, *P. aurelia*'s apparent equilibrium state could have been temporary. This is because, as already noted from the theoretical point of view, no single species can, in a competitive situation, reach an equilibrium state independent of the other. Then, logically, we cannot as yet dismiss the possibility that coexistence was on the way when Gause terminated his experiment. In other words, this iconic example of 'the competitive exclusion principle' is, strictly speaking, inconclusive.

Altogether, neither an apparent coexistence nor an apparent elimination as observed in an experiment may reject or support the notion of 'competitive exclusion principle'. In this respect, I remember Gause remarking (towards the end of chapter 1 in his *Struggle for Existence*) that: 'Mathematical investigations independent of experiments are of but small importance...'. It was a good point, considering that, about the time Gause was busy experimenting, ecological studies and mathematical theories were still in their infancy. However, after so many decades since then, especially in the advent of digital computers, the gap has been narrowed so much that the converse has become apparent: the significance of an experimental study cannot be properly assessed without an appropriate theoretical study. However, I must again emphasize that, for a mathematical model to be used as a logical basis (or as an analytical tool) for experimentation, it has to be built upon ecological first principles.

Keeping this problem of uncertainty in mind, let us examine the validity of the 'exclusion principle.'

9.5 Reconsideration of the 'Competitive Exclusion Principle'

Many ecologists have cast serious doubt on the validity of the 'principle'. An early and well-known critique was G. E. Hutchinson's (1961) article in which he contended that varieties of plankton coexisted while sharing the same nutrient resources.

In my opinion, to cast doubt on the validity of the notion, it need not require involved arguments. It is well known among entomologists that many parasitoid species attack a single host species. Ornithologists, professionals and keen (some fanatic) bird-watchers alike are well aware that many bird species feed on the same resources. Hunters and trappers are familiar with the fact that many predatory species share a prey species without apparent conflict. So, why has the notion of the exclusion principle persisted so stubbornly in ecology? I can think of several reasons.

One is to avoid a dilemma: accepting Gause's law as true on one hand but, on the other hand, finding no good reasons to ignore the ubiquitous cases of apparent coexistence in nature. A compromise is typically the idea of the segregation of ecological niches: 'No two species of similar ecology can occupy the same niche'. But is this really true? Besides the aforementioned observations by the ecologists and common knowledge among the laymen, my answer is 'No!', either empirically or theoretically.

I have observed that my great and blue tits (see Chapter 1) do have their own favourite sites (niches) to look for their food (mostly insects) but, more often than not, feed on the same species of insects that occur even in the same sites on the same individual tree. Right at this moment, I see at least a dozen different species of birds come to the feeder just outside my study: I see skirmishes or supplanting actions every now and then but, on the whole, they share the feed quite peacefully. Also, the spruce budworm (*Choristoneura fumiferana*, Lepidoptera: Tortricidae) — the subject material for the long-term study that my colleagues and I conducted (Royama et al., 2017) — hosts many species (we have so far found more than 35) of hymenopteran and dipteran parasitoids. Although many of them attack the host at its different developmental stages (micro-niches), there are always several species that attack the host at the same stage with no noticeable sign of segregation. However, an empirical assessment cannot be definitive. Thus, a theoretical consideration is in order.

Consider an extreme case: the two species have the set of parameters in model (9.4) that are identical in values, i.e. $(r_m, b, c) = (r'_m, b', c')$. That is, the two species cannot be more similar in ecology. However, their identicalness occurs in two situations. First, $b = b' < c = c'$. This situation complies with the stipulations for category (i) in Table 9.1, implying stable coexistence. Second, $b = b' \geq c = c'$. This complies with the stipulations for category (ii), including the transitional state between categories (i) and (ii), and hence implies the elimination of one species. Thus, theoretically, the two species of similar ecology may or may not

9.5 'Competitive Exclusion Principle' · 247

coexist. In other words, the assertion that 'No two species of similar ecology can coexist' has no firm basis either empirically or theoretically.

In recent years, some theoretical studies have postulated the possibility of coexistence and how it can happen. An example that has drawn my particular attention is the work of Edmunds *et al.* (2003). Referring to the apparent coexistence in the aforementioned classical experiments of Thomas Park (1948) on two *Tribolium* species, Edmunds *et al.*'s theory suggests in short that two species can coexist if their life-history specificities meet certain conditions.

The proponents of the exclusion principle might argue that this theoretical work depended on assuming a particular type of life history that would not be generally applicable. The proponents might also argue that, in Park's experiments, only one series among so many produced a case of apparent coexistence. In other words, the proponents might see these theoretical and experimental cases as just a few exceptions to the principle.

I have a different opinion: I see that these existing (documented) cases are likely to be either one of the four categories of the interaction processes posted in Table 9.1. But then, a hardcore proponent might still argue: why have so few cases of coexistence popped up in these experiments or in theoretical studies? I can think of a few probable reasons.

First, there have not been a large enough number of experiments to support or reject either view. Perhaps this is largely because it is technically not easy to carry out this sort of experiment. For one thing, while theoreticians can change the structure and parameter values of their models at will to explore all sorts of possible situations, the experimentalists do not have the liberty of selecting (changing) suitable materials. Usually, an experimenter has little choice but sticks to just one pair of species, each coming with life-history specificities of its own beyond the control of the experimenter. After all, Gause used just two *Paramecium* species, and Park, two *Tribolia* with a few different strains. I would not be surprised if the processes with these materials happened to be in a category other than (i) in Table 9.1.

I consider that the notion of 'competitive exclusion principle' has few firm (persuasive) bases either theoretically or empirically. More seriously, the notion tends to look at nature through too narrow a window to see only a partial picture, the elimination of one species. This partial view has, as I see it, led some ecologists to a compromise in terms of micro-niche segregation. Don't get me wrong: I am not saying that niche segregation is false. I think it is true in many circumstances, but I would not accept it as playing a dominant role in directing the way the species have come to occupy certain ecological niches as they do today.

On the other hand, if we abandon the old notion and accept an alternative idea that coexistence is the norm theoretically and empirically rather than exceptional, we would be led to a whole new era of ecological thoughts. So, I suggest that we stop fussing about the validity of the 'exclusion principle' but look positively at nature through different windows.

9.6 Alternative Ways of Viewing Nature

9.6.1 Formation of Species Complexes

Those arguments for or against the notion of the exclusion principle are based primarily on those laboratory experiments and theoretical models in which two species are reared (cultured) together on the given resource in total confinement. In a natural environment, this is rather exceptional. Normally, as already mentioned in Chapter 8, a species has its own repertoire of resources. For example, many (most likely the majority of) parasitoid species that attack spruce budworm have alternative hosts. Furthermore, the repertoires of alternative hosts, which several species of parasitoids have, overlap each other. That is, a given host species is shared among several parasitoid species, which form a parasitoid complex. In the meantime, several alternative host species make up a host complex.

Generally, the system of complexes is made up of species from different taxa. For example, the snowshoe hare is preyed upon by a complex of predators including coyotes, foxes, wolves, lynx, bobcats, and several species of raptors (to remind you again, I mean the birds of prey, like hawks, owls, etc., not those from Jurassic Park). Meanwhile, many species that are preyed upon, including hares, rabbits, rodents, and grouse, form a prey complex. Altogether, we usually find a system of these complexes. What is more, as already discussed in Chapter 8, each complex often behaves like a single unit such that a predator (parasitoid)-complex and its prey(host)-complex behave more or less like a single predator species and a single prey species, respectively. Nonetheless, these system complexes have certain important attributes that differ from interactions between a single predator species and a single prey species.

9.6.2 Formation of an Open System

The model processes or the experimental setups we have considered so far can be said to be *closed* in that the species concerned are confined in a space with no choice of alternative resources. Unlike those, the natural system of the complexes is *open* in that each member (be it an individual or as a species) has the freedom to move in and out of a complex, or to move from one system (of complexes) to another. But how would they determine when and where to move? The most likely factor is the *profitability* of a given niche, as I discussed in Chapter 1: the birds I studied tended to select those feeding sites according to the profit they can gain there at any point in time.

However, the profitability at a given site changes from time to time, as the availability of the prey (host) species at the site changes all the time. Moreover, the profitability of a site for a given predator (parasitoid) species depends on the presence of other species with which it competes or shares the resources. To deal with these problems, each species has several potentially profitable sites (repertoires of hunting sites to which the species has been well adapted) and optimally allots its time among these sites.

Altogether, the process of optimization plays a significant role in the selection of feeding niches, or more broadly speaking, in determining the course of evolution of species in selecting a set of micro-niches. On this basis, I make one more suggestion before moving on to Chapter 10 of this book.

9.7 Struggle for Existence vs Optimization of Profitability

Ever since the Darwinian conception of 'the struggle for existence', there seems to have been a sentiment among ecologists to see nature as being a rough and tough place where every creature must struggle continually against adversities. I do agree with the sentiment, but only partially. I think too much emphasis has been placed on this aspect of a species' life. I think that 'struggle' is a rather passive way for the species to deal with adversities. There has to be a positive aspect that directs the course of species evolution. If an individual (of a species) looks for something or some place to make its life more efficient so as to increase its fitness, such traits (propensities) would be more readily favoured by natural selection. I consider the optimization of profitability to be one of them.

Appendix 9A: How to Calculate x^{**} and y^{**}

At the equilibrium densities x^{**} and y^{**}, the rate of changes $R_t = R'_t = 0$. Thus, substituting x^{**} and y^{**} for x_t and y_t, respectively, as well as $R_t = R'_t = 0$, in (9.2), we have:

$$0 = R_m - cx^{**} - by^{**} \qquad (9A.1a)$$

$$0 = R'_m - c'y^{**} - b'x^{**}. \qquad (9A.1b)$$

Then, multiplying (9A.1a) by c' and (9A.1b) by b, and subtracting one from the other, we find:

$$c'R_m - c'cx^{**} = bR'_m - bb'x^{**}$$

and solving the above for x^{**}, we find:

$$x^{**} = (c'R_m - bR'_m)/(cc' - bb').$$

Similarly, (9A.1a) × b' − (9A.1b) × c yields:

$$(c'c - bb')y^{**} = cR'_m - b'R_m$$

and hence:

$$y^{**} = (cR'_m - b'R_m)/(cc' - bb').$$

Appendix 9B: Infeasibility of Category (v) in Table 9.1, Section 9.4

Let me use the symbol '⇒' to designate 'this means or implies'. Noting that each of the parameters (R_m, R'_m, b, b', c, c') is positive, the assumptions and their implications in this category are:

$$D = cc' - bb' > 0 \Rightarrow cc' > bb' \Rightarrow c'/b > b'/c; \qquad (9B.1)$$

$$N_x = c'R_m - bR'_m < 0 \Rightarrow c'R_m < bR'_m \Rightarrow c'/b < R'_m/R_m; \qquad (9B.2)$$

$$N_y = cR'_m - b'R_m < 0 \Rightarrow cR'_m < b'R_m \Rightarrow R'_m/R_m < b'/c. \qquad (9B.3)$$

The results in (9B.2) and (9B.3) yield $c'/b < R'_m/R_m < b'/c$, which contradicts the result in (9B.1). In other words, the stipulations $N_x < 0$ and $N_y < 0$ are incompatible with $D > 0$ in category (va). Likewise, in category (vb), $D < 0$ is incompatible with $N_x > 0$ and $N_y > 0$.

Appendix 9C: How to Incorporate the Effect of Random Exogenous Influences in the Model

The simplest way to do this is the following. First, generate two (50%) correlated series (ε and ζ) of independent, identically distributed random numbers (iid, normal with mean 0 and variance σ^2) as described in Appendix 8C, Chapter 8. Now, transform model (9.1) into natural logarithms as in (9.2), and add ε and ζ such that we have:

$$R_t = (R_m + \varepsilon_t) - by_t - cx_t$$
$$R'_t = (R'_m + \zeta_t) - b'x_t - c'y_t.$$

After having done simulations in the above forms, transform the results back into linear scale in Figures 9.9 and 9.10.

10 · *Observations, Analyses, and Interpretations: A Personal View through the Spruce Budworm Studies*

Generally speaking, an ecological study goes through the processes of observations, analyses, and interpretations of results, or in short the what-how-and-why sleuthing process. In this final chapter of the book, I talk about the thoughts that I have conceived through my experiences with the spruce budworm (*Choristoneura fumiferana*, Lepidoptera: Tortricidae) to share them with population ecologists at large.

During the 1950s, my apprenticeship years in Tokyo, a series of important papers on budworm population ecology was coming out from New Brunswick, Canada. These were the works by the group of entomologists led by R. F. Morris (most belonging to the Canadian Forestry Service in the Maritimes region, the present Atlantic Forestry Centre), and became well known worldwide as the Green River Project. I read the papers with great interest.

Quite unexpectedly, it so happened that, after the 10-year work on bird predation of insects in Japan and England, I was employed by the Canadian Forest Service and was assigned to the Green River Project. Initially, I was supposed to work on predation by birds and small mammals as a controlling agent of forest pest insects. But, rather, I chose to do a reanalysis of the original Green River data that had been stored in the basement of the laboratory for decades. My motivation to do the reanalysis was that despite the most innovative approach and enormous amount of observations, I thought the data analyses and interpretations by the earlier investigators, published in the 332-page monograph compiled and edited by Morris (1963), had left much to be desired in certain key aspects. Thus, after the reanalysis (Royama, 1984, 1992), I launched a new study with my younger colleagues and, after 14 years of intensive study in the field and the laboratory, we have acquired a much-needed body of information to understand budworm population processes (Royama *et al.*, 2017).

In the following, I talk about the thoughts that I conceived as antitheses of (or amendments to) those hypotheses that had been intuitively conceived by my predecessors, especially their conjectures on the formation of epicentres, spread of outbreaks, and dichotomy of endemic and epidemic populations. Although specific to this species, my thoughts to be given here would hopefully be of use for the further progress of animal ecology in general.

First, let me describe an outline of these budworm studies and the species' life cycle.

10.1 An Outline of the Spruce Budworm Studies

10.1.1 Life Cycle of the Species

Spruce budworm in eastern Canada has one generation per year. Eggs are laid in early July to early August in masses, each containing about 20 eggs attached to a needle of the host trees (the balsam fir, *Abies balsamea*, and the spruces, *Picea* spp.). The average fecundity of a female is about 200 (10 masses) of eggs: it can be estimated by the weight of a female pupa.

The eggs hatch in about 10 days. The first-instar larvae (L1) immediately disperse within the tree crown or beyond by wind, spin hibernacula in suitable sites (e.g. inside old staminate flower bracts, under bark, behind lichen, or anywhere within the stand, not necessarily on the host trees), moult to second instar (L2), and enter diapause to overwinter. The L2 emerge from the hibernacula in late April to early May, and immediately disperse again to find the foliage of the host trees.

Having settled on the foliage, the L2 mine into the needles of previous years to feed. After feeding inside a few of these needles, each L2 moults to L3 and moves out to feed on a newly developing bud on the periphery of the branch. As new shoots grow, a mature larva webs needles together around itself to form a 'feeding tunnel'. Movements within or beyond the tree crown rarely occur among larvae after the L3 stadium unless the needles (including those of previous years) have been exhausted on the branch or tree. We never saw beyond-tree dispersal in our own study: it was observed once in one plot during the Green River study, but was documented as an extreme and rare event (Morris, 1963). Pupation occurs in late June on the foliage where the L6 (final-instar larvae) have been feeding.

Adults eclose in about 10 days. The pupal exuviae left by the adults tend to remain on the foliage for a while. Therefore, the number of these exuviae estimates the number eclosed. After having laid a portion of their eggs, the majority of females (as well as males) disperse long distances on the wing (depending on weather conditions) from their natal sites to lay the remainder of their eggs elsewhere (Greenbank et al., 1980). Spruce budworm completes its life cycle within a tree crown, except for the comparatively short periods of dispersal at the L1, L2, and adult stadia. The sex ratio in adults is practically 1:1.

10.1.2 Outline of the Studies

The Green River Project was mainly carried out from 1945 to 1959, extended to 1972 for additional research, along the Green River in the north-west corner of the Province of New Brunswick, Canada, neighbouring the Province of Quebec and the State of Maine. One of its major objectives was to conduct the most innovative life-table study by determining the rate of change in population density from a given developmental stadium to the following, and analysing the factors that determined the rate. [Traditionally, population density is expressed in terms of the number of individuals (larvae, pupae, eggs, or egg masses) per unit of a foliated branch. For instance, with a balsam fir branch, it is the number per unit surface area in square meter.]

The problem with the Green River study was that sampling of the population was limited to only once every stadium, and little was known in between. In our later study from 1981 to 1995 in the Acadia Research Forest of the Atlantic Forestry Centre, we conducted daily sampling in the field and rearing of the sampled individuals at all stadia, except for the winter diapause and adult periods. Thus, we knew the causes of mortality exactly.

Now, let us look into the key findings in these studies. Let us first look at the earlier view of budworm outbreaks.

10.2 Earlier View of Outbreaks

As summarized in Morris (1963), the Green River Project had produced two key hypotheses: (i) the formation of an epicentre where an outbreak was considered to have originated, from which the infestation spread out to neighbouring areas and (ii) the existence of two distinct equilibrium states, one at an endemic level and the other at an epidemic level.

After having gone systematically through the original data from the Project (which took me a full decade), I found neither of the two hypotheses was substantiated. The reason behind this is as follows. By the beginning of the 1980s, a large body of information on annual changes in budworm populations (in terms of egg-mass density) since the early 1950s had become available from the long-term pest-control program across the Province of New Brunswick. To see the dynamical pattern of budworm populations, I divided the entire area of the Province into 18 (66×66 km^2) blocks, identified with alphabetical letters from north to south and numerically from west to east. In each block, the average egg-mass density is graphically plotted after a logarithmic transformation. The results (from 1952 to 1996) are graphed in figures 2 and 3 of Royama et al. (2005), and reproduced here in Figure 10.1.

The graphs show no clear sign of epicentres from which infestations spread to the surrounding areas. Neither do the graphs show the dichotomy of endemic–epidemic equilibrium states. What we actually see instead is a continuously cycling budworm population everywhere across the Province, and the cycle was found in our later study (Royama et al., 2017) to be attributable to the interaction between a host complex (in which the budworm is the major constituent) and its natural enemy complex, comprised largely of insect parasitoids, invertebrate and vertebrate predators and, to a much lesser extent, a few pathogenic microbes.

After all, the hypothetical existence of epicentres and double-equilibrium states turned out to be unsubstantiated. Having recognized this, a question arises: how did my predecessors conceive these hypotheses? The answer is basically: they were pioneers and did not have as much information as I had as a second-generation investigator. However, the important point here is that I learned a lot from finding how they were misled, and I feel I should share what I found with the ecologists at large. So, let me examine how the pioneers arrived at the epicentre hypothesis and the spread of infestation.

10.2.1 Origin of the Epicentre Hypothesis

The most likely source of the hypothesis was the so-called infestation maps, which were used for detecting 'hot spots' in the woods and for deciding where to apply insecticide aerially. On a temporal series of these maps, you would first see a few spots where some damage to the foliage of the host trees became evident in aerial photographs in conjunction with ground surveys. These spots were considered to be epicentres.

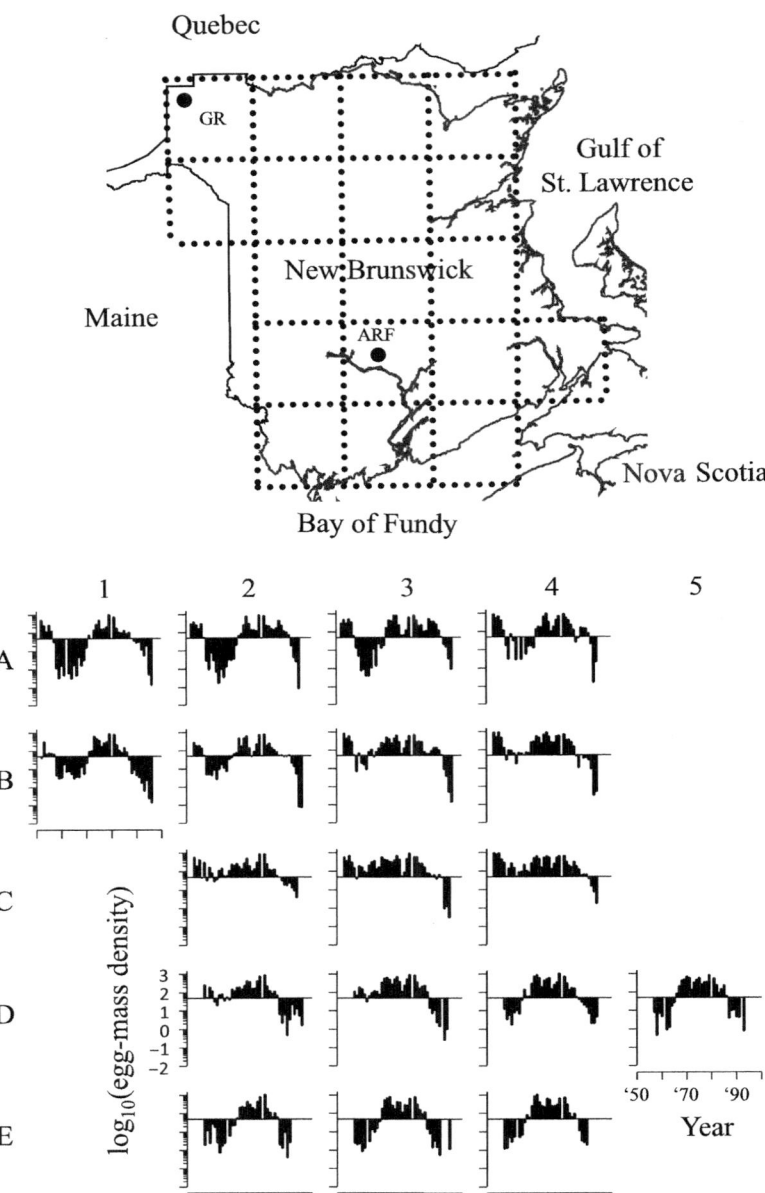

Figure 10.1 Province of New Brunswick and outbreak cycles of budworm populations as depicted by the egg-mass counts. GR, Green River Field Laboratory in block A1. ARS, Acadia Research Forest in block D3. After figures 2 and 3 in Royama *et al.* (2005). (With permission from John Wiley & Sons, © 2005 by the Ecological Society of America)

Then, the patches began to grow in size over subsequent years to finally cover an extensive part of the Province, the situation recognized as the 'outbreak state'. However, after a while, these heavily infested areas began to recede to an invisible level, and no areas of infestation became noticeable for a comparatively long period of time (recognized as the 'endemic state') until a new cycle began.

Now, the above perception of the outbreak cycle appears to be supported by actual measurements of population levels estimated by sampling. Plotting these estimates against the years in Figure 10.2a looks as though a local budworm population stays at an endemic level for a long period of time but every so often suddenly rises to an outbreak level: figure 9.20 in Royama (1992) gives a few more examples across eastern

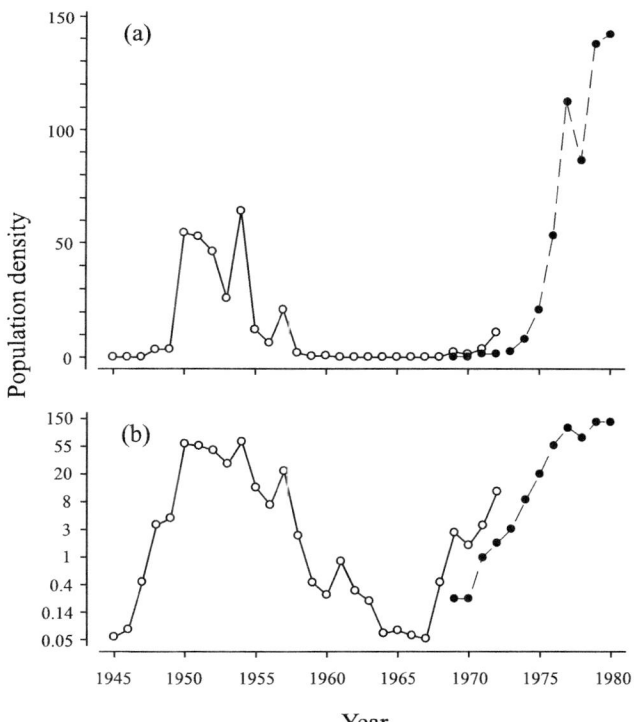

Figure 10.2 Budworm outbreak cycles as depicted in the Green River study. Open and solid circles are (3rd to 4th instar) larval and egg-mass densities, plotted in linear scale in (a) and in logarithmic scale in (b). After figures 1 and 30 in Royama (1984). (With permission from John Wiley & Sons, © 1984 by the Ecological Society of America)

Canada. But this method of graphing the observed data creates a problem.

As pointed out in Chapter 3, an organismic population is a positive entity, and plotting it against time often fools the viewers. This is because changes in the population at a very low level can be pressed so much against the time (horizontal) axis where the details could become imperceptible. The problem can be resolved by a logarithmic transformation. Indeed, Figure 10.2b, a log-transformed version of Figure 10.2a, shows no sign of the budworm populations staying at a low level for any prolonged period of time; the population continually cycled without a pause at any moment, as expected in an interaction between a host and the natural-enemy complex, as discussed in Chapter 8.

Having found the origin of the problems in the earlier hypotheses, we should look for alternative ways to interpret the infestation maps. My alternative is quite straightforward. First, what had been considered to be an epicentre was one of those areas where the local budworm population happened to have reached (slightly ahead of the neighbouring areas) a certain level at which the signs of defoliation became so visible. Second, in the neighbourhood of the spot with the first sign of defoliation, all other local populations were increasing more or less simultaneously, and they quickly caught up to show signs of defoliation there as well. After all, on the infestation maps, it looks as though an outbreak is 'spreading' from an apparent 'epicentre', which turned out to be just an illusion.

I now look at the endemic–epidemic dichotomy a little more closely. Apparently dissatisfied with the mere intuitive perception of dichotomy, Morris tried to justify it by putting forward a hypothesis of the underlying mechanism in the form of a graphical model.

10.2.2 Hypothetical Mechanism Underlying the Endemic–Epidemic Dichotomy

In Chapter 4, I introduced the method of reproduction curves that Ricker (1954) used extensively in his analysis of fish populations, which inspired Morris (1963) to apply the method to his analysis of spruce budworm.

Recall from Chapter 4 that the Ricker reproduction curve in its original format was to plot population density x_{t+1} against x_t, which graphically generated the dynamical pattern of the population series $\{x_t\}$. Morris modified this format into the plot of $X_{t+1} = \log(x_{t+1})$ against

10.2 Earlier View of Outbreaks · 259

$X_t = \log(x_t)$, and I further modified it to the plot of $R_t = \ln(x_{t+1}/x_t)$ against $X_t = \ln(x_t)$. The three formats carry exactly the same information, but I use my format because it exhibits the simplest form and is easy to interpret. Thus, using theoretical model (6.1) from Chapter 6, I generated a variety of reproduction curves in Figures 6.1 and 6.2. Now, notice that every curve in these figures decreases monotonically as population increases. Notice in particular that, as it decreases from above the horizontal axis on which $R_t = 0$, the curve crosses the axis only once: the point of the intersection defines the equilibrium density of the population, marked X^* in Figure 6.1.

Being inspired by Ricker, however, Morris envisaged that a reproduction curve could be sinusoidal, or wavy, such that the curve may cross the $(R_t = 0)$-axis more than once. To visualize this attribute, consider curve (A) in Figure 6.1. Consider in particular that, as it continues to decrease after the point X^*, the curve begins to increase to cross the $(R_t = 0)$-axis from below, but after a while it begins to decrease again to cross the axis from above, creating a second (upper) equilibrium point. Figure 10.3 schematically represents the situation, in which X^+ and X^{++} designate the lower and upper equilibrium densities, respectively.

The ecological significance, as Morris envisaged, of the sinusoidal curve is the following. The point where the increasing section of the curve crosses the $(R_t = 0)$-axis from below (designated by X^o in Figure 10.3) is an equilibrium point, but it is unstable: if the population deviates from it to the left, it decreases towards the lower equilibrium

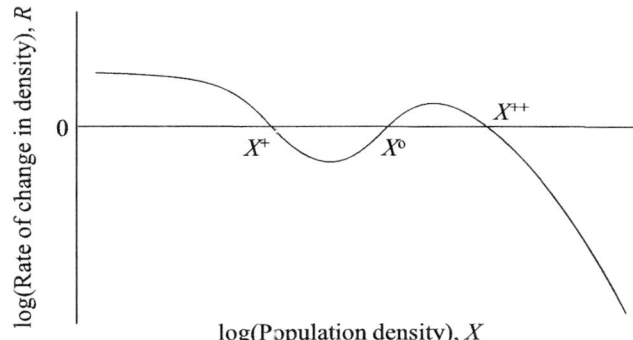

Figure 10.3 Sinusoidal reproduction curve as hypothetical mechanism to represent the dichotomy of endemic and epidemic equilibria. X^+: endemic equilibrium point. X^{++}: epidemic equilibrium point. X^o: release point.

point; if deviating to the right, the population increases towards the upper equilibrium point.

Morris considered that a budworm population would usually stay about the lower equilibrium point X^+; but under a favourable climatic condition, it could increase past the middle (unstable) equilibrium point $X°$, which he called a 'release point'; and the population would head towards the upper equilibrium point X^{++}. However, after a while, excessive defoliation would reduce the survival rate and the population would return to the lower equilibrium state and would remain there until the beginning of another outbreak cycle. Although this idea appeared to have appealed to his contemporaries, I found it problematic for two reasons.

First, how can a reproduction curve change its direction from decreasing to increasing at a certain population density? It is a physical possibility in that a curve of such an attribute can be readily created (represented) mathematically, e.g. a polynomial equation of a certain degree. The problem is, however, that I am unable to find an ecologically feasible mechanism that can generate a sinusoidal reproduction curve by any means. [Try it if you can, remembering that a reproduction curve is an endogenous process, i.e. x_{t+1} as a function of x_t, under no exogenous influences.]

Second, having recognized through my own experience that the budworm dynamics is basically a host–natural enemy complex interaction, it cannot be represented by a single reproduction curve but by a surface as illustrated in Figure 8.7 (i.e. Surface X for prey) in Chapter 8. The following is my understanding of how Morris envisaged (or rather was misled by) the sinusoidal reproduction curve.

Recall that the three-dimensional Surface X can be projected (after a log-transformation) onto a two-dimensional plane by plotting the rate of change R_t against the population density X_t with a family of conditional reproduction curves, each corresponding to a given particular location on the Y (predator) axis, as illustrated in Figure 8.8. Also, take a look at Figure 8.10 with simulated data points that form an elliptical orbit. This graph must have been in effect what Morris saw with his observed data: 'must have ... in effect' in that he did not provide a graph with actual data points.

Now, let us use a little imagination with the help of Figure 8.10. Suppose that many study plots in the Green River Project were complete with frequent observations of several different population cycles, and R_t was plotted against X_t. Then, Morris should have seen as many sets (as the number of the study plots) of data points, each set forming an orbit about its single equilibrium point X^{**} (x^{**} being given in Appendix 8B), and all orbits lining up along the ($R_t = 0$)-axis in the order of their equilibrium points.

However, the Project did not have enough data points from a single study plot to draw a reproduction curve with an adequate length. Thus, Morris pooled data from many study plots, most of them having been observed only for a fraction of a cycle. Consequently, when the observed X_{t+1} was plotted against X_t (equivalent to the plot of R_t against X_t), he saw a scattergram of data points along the 45° diagonal reference line (equivalent to the horizontal $R_t = 0$ line), some points lying above, and others below, the line. Then, I can readily imagine that Morris saw an apparently sinusoidal pattern in the scattergram through which he drew a regression curve by hand, as in fact shown in his figure 18.3 (Morris, 1963, p. 126). Thus, after all, the idea of a sinusoidal reproduction curve turns out to be deceptive, neither conceivable theoretically nor supported empirically.

Now, let us move on to the question: why does a local budworm population reach an outbreak level so often?

10.2.3 Cause of an Outbreak

Before delving into the theme, let me make clear what is meant by 'outbreak'. This is an expression, used among the pest-control people, to describe the level of budworm population over which the loss of timber becomes economically intolerable. It is not an ecologically definable measure of population level. Nonetheless, it is a convenient expression for describing an extreme (extraordinary) situation.

As far as I see, the cause of a budworm outbreak (in the above sense) is quite simple in principle. First, the intensity of competition within a local budworm population is comparatively low so as to potentially allow the population to reach an extremely high level. Second, the effect of natural enemy complex (parasitoids, predators, and pathogens) during each budworm generation is comparatively small so as to allow the budworm population to keep increasing slowly over many generations.

Spruce budworm clearly has the first attribute. As already mentioned in Chapter 8, its larvae tolerate an extreme level of crowding: as many as 100 larvae might be found feeding together on just one mid-crown branch of mature balsam fir without a sign of conflict. But how about the second attribute: the effect of the natural-enemy complex?

The most prevalent pathogens are the unicellular parasitic fungi (Microsporidia), which are known to be low in virulence. Other pathogenic fungi, bacteria and viruses are always present. However, as far as I am aware, a level of infection with these pathogens high enough to suppress a local budworm population to a low level has been reported only infrequently throughout the history of budworm studies in northeastern North America. Predation by vertebrate (birds) and arthropods (spiders, pentatomid bugs and carabid beetles) were regularly observed in our own study, but their effect was assessed as being ineffective for preventing the budworm population from increasing. So, let us look at the remaining major mortality factor, the complex of parasitoids. Evidently, their collective effect during each generation is on average rather small, allowing the budworm population to increase slowly to reach an outbreak level. I can think of a good reason why this happens.

As already mentioned in Chapter 9, our study found at least 35 different species of parasitoids, each attacking budworm at a particular one of its developmental (egg to pupal) stadia. However, we also found at least 18 of these primary parasitoids being attacked by at least 22 species of hymenopteran hyperparasitoids (cf. appendix S11 in Royama *et al.*, 2017): I would not be surprised if all of the primary parasitoids were found to be attacked by hyperparasitoids. Thus, the presence of hyperparasitism must certainly have reduced the potential of the primary parasitoids to control the budworm population, so as to allow it to increase to an extreme level. Little is known about the ecology of hyperparasitism, and what I have conceived above remains a conjecture. I leave the task of testing it as a major subject of the future studies by the next generation of investigators.

Now, a big question remains to be answered: how do budworm outbreaks occur simultaneously all across the Province of New Brunswick? Phenomenologically, this is a consequence of all local populations cycling more or less in unison. Thus, our interest is to find a plausible mechanism that tends to bring these local population cycles into synchrony.

10.2.4 Population Synchrony and a Myth of Moth Dispersal

Among those engaged in forest pest-control programmes, it is very well known that spruce budworm moths of both genders are strong fliers, and that, in particular, female moths disperse while still at least partially gravid. Thus, it has been naturally assumed that aerial dispersal of the moths would bring the local budworm cycles into synchrony. However, for this intuitive idea to be plausible, the following two premises must hold true: (i) moth dispersal is non-directional like a diffusion process in chemistry and (ii) the trend of the population cycle at the locality where the dispersing moths land is affected by the additional eggs laid by these invading moths. My reanalysis of the Green River data and the subsequent study of moth dispersal support neither of the premises. Let us look at the first premise: moth dispersal is a diffusion process.

During the 1970s, the Atlantic Forestry Centre launched an intensive study of moth dispersal by radar (Greenbank et al., 1980). This revealed that, unlike a diffusion process, moth dispersal tended to be directional: once in the air, they flew with the prevailing wind which tended to be directional, depending on the climatic conditions of the day. Also, a prevailing wind current could act as a funnel, collecting those moths taking off from various locations all along the wind path and dumping all of them in a single spot: this creates the phenomenon known as a mass invasion of moths as reported by local newspapers every now and then. Evidently, moth dispersal is unlikely to be similar to a diffusion process of chemical substances. Thus, Premise (i) does not hold true.

Now, let us take a look at Premise (ii): the effect of the additional eggs laid by the invading moths on the local population cycle. Let us take a look first at the effect of invasion. An invasion of moths at a given locality is readily detectable by estimating the ratio between the total number of eggs (E) found on the foliage and the total number of moths (both sexes) (M) locally emerged (empty pupal cases left also on the foliage), or the E/M ratio in acronym. If no dispersal (neither immigration nor emigration) occurred, then the ratio should be more or less equal to one half of the mean fecundity of a local female moth: 'one half' because the sex ratio in budworm is on average 1:1. [For fecundity, see Section 10.1.1 on the Life cycle.]

Figure 10.4 shows annual changes in the E/M ratio in comparison with half of the fecundity (of the natal females) marked by the dashed line (call it the F/2 line) at two study plots (named G4 and K1) in the Green River Project, carried out within block A1 of Figure 10.1. At plot G4, the ratios were substantially larger in a few years than the F/2 line, indicating moth invasions. On the other hand, in another few years, the ratios were distinctly lower than the F/2 line, indicating emigrations of the locally raised moths. In contrast, at plot K1, the ratios were mostly less than the F/2 line, indicating emigrations of the local moths in most years. Clearly, then, an E/M ratio is a balance between the gain (due to immigration of foreign moths) and the loss (due to emigration of local

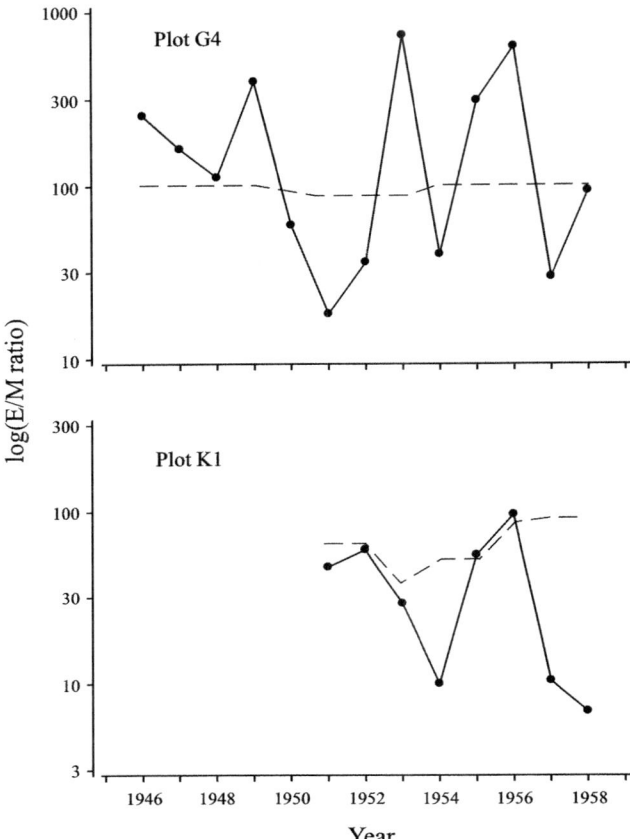

Figure 10.4 The variations in the E/M ratio observed at the Green River study plots G4 and K1. After figure 10 in Royama (1984). (With permission from John Wiley & Sons, © 1984 by the Ecological Society of America)

moths): a ratio higher (or lower) than the F/2 line indicates a net invasion (or emigration).

Thus, intuitively, it would be natural to assume that the effect of moth invasion must have been high in plot G4 but low in plot K1. In particular, those substantial invasions that occurred in plot G4 would have kept the population there increasing. [Incidentally, the average population level was so low in plot G4 that no noticeable defoliation occurred, whereas the population in plot K1 was high enough to cause severe defoliation.] However, what actually happened, as illustrated in Figure 10.5, was counterintuitive.

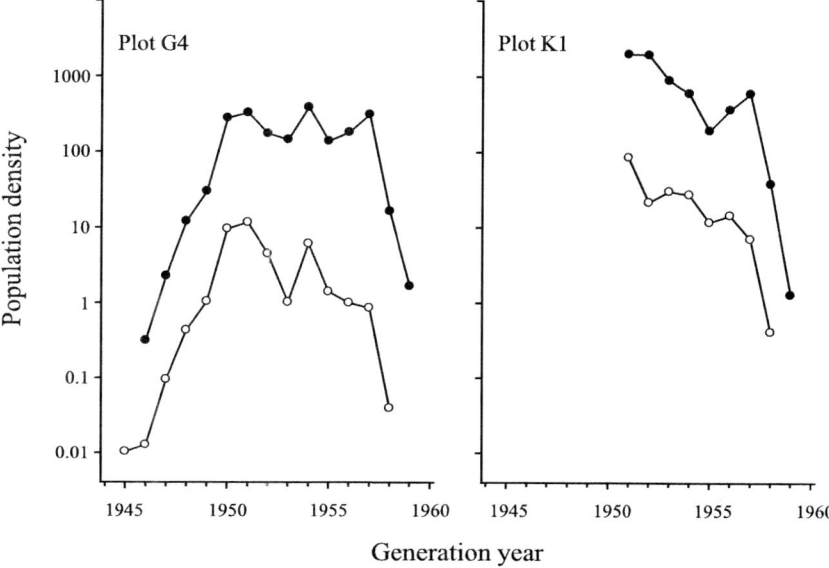

Figure 10.5 Annual changes in budworm population density as observed in Green River Project at the two study plots of Figure 10.4. Solid and open circles are eggs (individual) and pupal densities, respectively. After figures 4 and 6 in Royama (1984). (With permission from John Wiley & Sons, © 1984 by the Ecological Society of America)

In both plots, the populations declined at more or less the same time and in a similar manner, regardless of the differences in the average level of E/M ratios. Even substantially high E/M ratios in plot G4 did not reverse the trend. More precisely, in either plot, although the egg density increased temporarily following a comparatively high E/M ratio, by the time all the larvae had pupated, the population level returned to the level that would have been expected to be realized without moth invasions. Presumably, as our later study (Royama et al., 2017) suggested, this was due to the constituents of the local parasitoid complex sensitively responding to the temporary increase or decrease in the host population at the developmental stadia of their preferences. In other words, the sensitive responses of the parasitoid complex nullified the effect of the variation in the E/M ratio.

Altogether, I reasoned that moth dispersal was unlikely to be a direct cause of synchrony in the budworm population cycles, but what else could be the cause? The most plausible mechanism that I can think of is the following.

10.2.5 The Moran Effect

In the late 1960s, while studying models of predation and parasitism, I came across an interesting paper by P. A. P. (Pat) Moran (1953) on the synchrony of the Canada lynx (*Lynx canadensis*): as I reviewed in detail (Royama, 1992), the well-known lynx 10-year cycles are remarkably well synchronized all along the Canadian boreal forest zone. Pat explained it in terms of what I later coined 'the Moran effect', a theorem based on the linear time-series model. The theorem as applied to budworm is in short as follows. 'If several local populations are independently cycling due to the interaction with the local natural-enemy complex, the cycles will be brought into synchrony under some exogenous influences, if the influences are correlated among the localities.'

Now, the effect of moth dispersal (as measured by the E/M ratio) may act like an exogenous influence. Then, a correlation in the E/M ratio among the local populations should bring them into synchrony. The key here is the word 'correlation'. It does not matter if the ratio at a locality deviates above or below the corresponding F/2 line. What matters is the correlation in the pattern of variation in the ratio between the local populations, as it appears (in that it is not conclusive) to be the case between the study plots G4 and K1 in Figure 10.4: a few more examples are given in figure 10 in Royama (1984). So, the question here is: what can actually cause the correlation in the E/M ratio? My tentative answer (conjecture) is the following.

As already mentioned, the E/M ratio in a given locality is a balance between the gain of eggs brought in by the foreign moths and the loss of eggs carried out by the locally raised moths. Now that the gain from outside has been found to be localized, it is unlikely to be correlated across much wider areas of the Province. On the other hand, as I conjectured in Royama *et al.* (2017), the emigration of moths is most likely to be universal and, hence, is likely to be correlated under the influence of prevailing climatic conditions: good weather encourages moths to take off, whereas bad weather discourages them to do so. This is the plausible mechanism of synchrony in the budworm outbreak cycles that I have so far conceived. As already mentioned, it is a conjecture, and its likelihood has to be evaluated by some means, a challenging task for the active investigators.

The above arguments show that the processes of observations, analyses and interpretations are integrated parts of an ecological study in that, as we make observations, we analyse and interpret what we have seen, and continue to do the same repeatedly as the research progresses. Nonetheless, each process has its own specific way to be done and may create its own specific problem, as I consider in the following.

10.3 Thoughts on the Basic Processes of Ecological Studies

10.3.1 Observations

Usually, observation is the first stage of ecological field studies. When venturing into a previously unexplored area, we naturally attempt to observe as many aspects as we can manage physically and financially. This was what the early investigators of the Green River Project attempted: they selected a large number of study plots to include as many different stand types (species composition and age) as they could handle. This approach enabled the investigators to see differences, which guided them to further investigating what had caused the differences. The problem was, however, that they did this at the expense of sampling frequency at each study plots: at most only once every developmental stadium of budworm, and that was too sporadic and some details were missed.

After having carefully studied their observations, I saw that the general trend in the observed population changes was much the same, regardless of stand types. This led me to launch a new study in which I limited observations in just one or two study plots at a time to carry out daily sampling in the field and rearing the samples in the laboratory. This scheme enabled us to determine changes in the budworm survivorship as a continual process throughout its developmental period.

In the above examples, intensive observations over two generations of investigators yielded two different perceptions: one was to see (or to place emphasis on) dissimilarity and the other, similarity. However, the two perceptions differed merely in their own emphases that were mutually compensatory and the two together provided deeper insight into the budworm population processes.

As for the way that the endemic–epidemic dichotomy was conceived in the Green River Project, the nature of the problem was altogether different. The investigators were misled when they tried to visualize the observed trend in population in a graphical form: plotting the observed numerical data in a non-negative linear scale. Although a natural way of drawing the graph, it had misled many people (e.g. those advocating the sigmoidality of the logistic law) during the early stage of developing population ecology, a point we must bear in mind in preparing graphs of the observed data for analysis.

Altogether, we should recognize that observations may require several generations of researchers to collect a reasonable amount of reliable information.

10.3.2 Analysis of Observation

In my apprenticeship years, I was taught that speculations were unacceptable in writing an ecological paper. However, I remember that a prominent biologist pointed out that a speculation is permissible provided that, one step behind it, there is a supporting fact. I think this is a good point. I even think that a conjecture is necessary, especially at the beginning of a new study: we only need to reject or accept it for further progress of the study. Nonetheless, a question may arise: what could be a supporting *fact*? My perception is as follows.

Usually, a fact is an empirically observed event that has been known or proven to be true, but I expand it to a theoretically conceived event that has been proven to be true. For instance, the Pythagorean Theorem of the right-angled triangle is a theoretical fact, and so are all proven theorems in mathematics in general. This expansion is necessary because a theoretical consideration, especially in the form of a theoretical model, is an indispensable tool for the analysis and interpretation of what has been observed. However, we should keep in mind a few fundamental issues. First of all, there are two major categories of models in ecology: descriptive and analytical.

A descriptive model may be built primarily for reproducing (or mimicking) an observed dynamical pattern, chiefly in the form of computer model such as those built for prediction purposes in pest-management programmes. We must use caution, however. Because its primary purpose is to mimic the observed pattern, it is often comprised of multiple layers of assumptions, some of which are almost inevitably speculative. The real danger is that some models are based on assumptions that are ecologically infeasible. Nonetheless, the computer would output the result for display, which may be put into practical use uncritically. After all, we should keep in mind that ecologically infeasible models may create more myths than revealing truths.

In contrast, an analytical model is used for revealing the mechanism underlying the observed pattern. To serve this purpose, the model must meet the following prerequisite. It should be built upon ecological first principles, i.e. upon those attributes that are undoubtedly true or axiomatic. Then, provided that the inference that follows is logically correct, we can rely on the resultant model as true. But it should be noticed that the model would play two roles.

One role is to provide theoretical criteria on which we interpret the observed pattern. If the model fitted the observation, so be it, although this would not happen too often. The pragmatic value of this role is that the model would become a useful tool when, paradoxically, it did not fit the observation: a careful consideration of the reason why it did not fit would often provide insight.

The other role that an analytical model plays is the following. We must recognize that not all ecological attributes are observable empirically. For instance, as discussed in Chapter 9, the possibility of the coexistence of two species cannot be demonstrated experimentally. An experiment can only confirm the elimination of one species, whereas an apparent coexistence would be inconclusive because the length of an observation is limited in general. This is where a well-constructed analytical model becomes indispensable: it provides a theoretical fact, the only source of information which enables us to perceive unobservable attributes.

10.3.3 Interpretation of Results

An oft-used method of interpreting (understanding) the nature of an observed event is to use an analogy with 'something' that are already familiar with. For instance, the physics of the electric current can be understood by analogy with water currents. In ecology, an analytical model may be used as this 'something' to understand an observed event. More precisely, doing simulations with the model by varying its parameter values would make us familiar with what would be expected under a specific set of conditions.

The success of this approach would depend on how we choose an appropriate model. To do this, however, we need to understand the nature of the model precisely. In particular, we need to know how the model is constructed on what assumptions. This is where we require knowledge from theoretical ecology, and to make use of it, we must understand the mathematics behind the model, which in turn implies that we must acquire adequate mathematical skills. Otherwise, there will not be as much progress in population ecology as we need to make to cope with the difficulties that we would likely face in the rapidly changing environments.

10.4 Concluding Remarks in the Quest for Certitude

A population process is a stochastic process full of unpredictability and uncertainty, resulting in our observation of it imprecise. However, the way we deal with this problem in thinking can, and should, be precise. This is exactly where mathematics becomes an absolute necessity. However, in practice, there is a limit to the extent we the ecologists acquire adequate mathematical skills, and close cooperation with mathematicians becomes desirable.

Conventionally, there appear to be a demarcation between empirical ecology and theoretical ecology. It used to be a common practice that an ecologist takes a pile of numerical data to a theorist and asks them to analyse it; the theorist does a statistical analysis, or builds a model out of it if asked; and the ecologist accepts the result as given. It might work, but more often than not it fails. A major problem lies in inadequate communications between the two sides of the demarcation.

As far as I can see, there is only one way to eliminate the gap: the empirical ecologists must acquire mathematical skills so that the products of theoretical ecology are correctly understood and appropriately incorporated into the process of interpretation.

Nonetheless, owing to our limitation in acquiring adequate skills, this may not be easily achievable. I have a solution: I have always had a personal adviser in mathematics and statistics in the local university. Whenever I had a problem, I went to see him for advice. I never tried to explain what my problem was in ecological terms but in mathematical forms. In other words, I asked the mathematician a mathematical question: how to incorporate the answer into the ecological problem was my task. This effectively avoided the problem of miscommunication. So, I strongly recommend that empirical ecologists learn at least basic (elementary) mathematics and find a personal mathematical adviser in a nearby institution.

If this approach becomes a general trend, the future of population ecology should be bright, and I hope my philosophy presented in this book will help the trend to grow.

References

Adrewartha, H.G. and Birch, L.C. (1954) *The Distribution and Abundance of Animals.* University of Chicago Press, Chicago.

Brännström, Å. and Sumpter, D.J.T. (2005) The role of competition and clustering in population dynamics. *Proceedings of the Royal Society B*, **272**: 2065–2072.

Chapman, R.N. (1931) *Animal Ecology: With Special Reference to Insects.* McGraw-Hill, NY.

Edmunds, J., Cushing, J.M., Costantino, R.F., Henson, S.M., Dennis, B. and Desharnais, R.A. (2003) Park's *Tribolium* competition experiments: a non-equilibrium species coexistence hypothesis. *Journal of Animal Ecology*, **72**: 703–712.

Feller, W. (1940) On the logistic law of growth and its empirical verifications in biology. *Acta Biotheoretica*, **5**: 51–66.

Gause, G.F. (1934) *The Struggle for Existence.* Williams and Wilkins, Baltimore.

Greenbank, D.O., Shaefer, G.W., and Rainey, R.C. (1980) Spruce budworm (Lepidoptera: Totricidae) moth flight and dispersal: new understanding from canopy observations, radar, and aircraft. *Entomological Society of Canada Memoir* **110**.

Hassell, M.P. (1975) Density-dependence in single-species populations. *Journal of Animal Ecology*, **44**: 283–295.

Holling, C.S. (1959) Some characteristics of simple types of predation and parasitism. *Canadian Entomologist*, **91**: 385–398.

Howard, L.O. and Fiske, W.F. (1911) Importation into the United States of the parasites of the gypsy moth and the brown-tail moth. US Department of Agriculture, Bureau of Entomology, Bulletin No. 91: 311 pp.

Hutchinson, G.E. (1961) The paradox of the plankton. *American Naturalists*, **95**: 137–145.

Keith, L.B. (1963) *Wildlife's Ten-year Cycle.* University of Wisconsin Press, Madison.

Kettlewell, H.B.D. (1958) Industrial melanism in the Lepidoptera and its contribution to our knowledge of evolution. *Proceedings of the Xth International Congress of Entomology*, **2**: 831–841.

Lack, D. (1954) *The Natural Regulation of Animal Numbers.* Clarendon Press, Oxford.

Leslie, P.H. (1948) The use of matrices in certain population mathematics. *Biometrika*, **33**: 183–212.

Leslie, P.H., Park, T., and Mertz, D.B. (1968) The effect of varying the initial numbers on the outcome of competition between two *Tribolium* species. *Journal of Animal Ecology*, **37**: 9–23.

Lotka, A.J. (1925) *Elements of Physical Biology*. Williams and Wilkins, Baltimore.
Moran, P.A.P. (1950) Some remarks on population dynamics. *Biometrics*, **6**: 250–258.
Moran, P.A.P. (1953) The statistical analysis of the Canadian lynx cycle. II. Synchronization and meteorology. *Australian Journal of Zoology*, **1**: 291–298.
Morris, R.F. (Ed.) (1963) The dynamics of epidemic spruce budworm populations. *Memoirs of the Entomological Society of Canada*, **31**: 311–320.
Nicholson, A.J. (1954) An outline of the dynamics of animal populations. *Australian Journal of Zoology*, **2**: 9–65.
Nicholson, A.J. and Bailey, V.A. (1935) The balance of animal populations.—Part I. *Proceedings of the Zoological Society of London*, **3**: 551–598.
Park, T. (1948) Experimental studies of interspecies competition. I. Competition between populations of the flour beetles *Tribolium confusum* Duval and *Tribolium castaneum* Herbst. *Ecological Monographs*, **18**: 265–308.
Pearl, R. and Reed, L.J. (1920) On the rate of growth of the population of the United States since 1790 and its mathematical representation. *Proceedings of the Academy of Sciences*, **6**: 275–288.
Ricker, W.E. (1954) Stock and recruitment. *Journal of the Fisheries Research Board of Canada* (forerunner of the current *Canadian* Journal of Fisheries and Aquatic Sciences), **11**: 559–623.
Royama, T. (1966) Factors governing feeding rate, food requirement and brood-size of nestling Great Tits *Parus major*. *Ibis*, **18**: 313–347.
Royama, T. (1970) Factors governing the hunting behaviour and selection of food by the Great Tit (*Parus major* L.). *Journal of Animal Ecology*, **39**: 619–668.
Royama, T. (1984) Population dynamics of the spruce budworm *Choristoneura fumiferana*. *Ecological Monographs*, **54**: 429–462.
Royama, T. (1992) *Analytical Population Dynamics*. Chapman and Hall, London.
Royama, T., MacKinnon, W.E., Kettela, E.G., Carter, N.E., and Hartling, L.K. (2005) Analysis of spruce budworm outbreak cycles in New Brunswick, Canada, since 1952. *Ecology*, **86**: 1212–1224.
Royama, T., Eveleigh, E.S., Morin, J.R.B., Pollock, S.J., McCarthy, P.C., McDougall, G.A., and Lucarotti, C.J. (2017) Mechanisms underlying spruce budworm outbreak processes as elucidated by a 14-year study in New Brunswick, Canada. *Ecological Monographs*, **87**: 600–631.
Ruiter, L. de (1952) Some experiments on the camouflage of stick caterpillars. *Behaviour*, **4**: 222–233.
Smith, H.S. (1935) The role of biotic factors in the determination of population densities. *Journal of Economic Entomology*, **28**: 873–898.
Tinbergen, L. (1960) The natural control of insects in pinewoods. I. Factors influencing the intensity of predation by songbirds. *Archives Néerlandaises de Zoologie*, **13**: 265–336.
Verhulst, P.-F. (1838) Notice sur la loi que la population suit dans son accroissement. *Correspondence Mathématique et Physique*, **10**: 113–121.
Verhulst, P.-F. (1845) Recherches mathématiques sur la loi d'accroissement de la population. *Mémoires de l'Académie Royal de Bruxelles*, **18**: 1–38.
Williamson, M.H. (1970) *The Analysis of Biological Populations*. Edward Arnold, London.

Index

accumulation of errors, *see* random walk
age class, 58, 60, 136, 146–7
aggregation, 107, 120–1, 126, 129, 137
analytical model, 268–9
apex of curve, *see* tracing of curves
autocorrelation, 161, 175, 201–2
  function (ACF), 180, 201
autocovariance, 147, 149, 159, 161, 164–6, 178
  function (ACVF), 166, 180
averages
  ensemble, 165
  time, 165
azuki bean weevil, *see* Utida's experiments

binomial
  coefficient, 103–4, 115
  distribution, *see* probability distribution
  series, 102
  trial, 100, 103
biotic-factor theory (school), *see* regulation of populations
birth and death, 35, 181
Brännström–Sumpter model, 114
breeding season, 35, 39, 139, 181
Brownian motion, 156

carrying capacity (of environment), 38, 48–9, 91
census data, 33, 37, 71–4, 90
chaos (deterministic), 81–2, 84
climate changes, 86, 88, 154
climatic-control theory (school), *see* regulation of populations
climatic influences, 145, 171–2
closed process, 249
coexistence (of competing species), 228–32, 238, 241, 244–8, 269

common version logistic model, *see* logistic models
competition
  geometric model of, 40
  intensity of (in), 42, 47, 49, 61, 63, 98, 105, 121, 127–8, 226, 261
  interspecific, 225–51
  intraspecific (within-population), 32, 40, 63, 85, 108, 113, 148, 183, 233
  scramble and contest, 120, 125–6, 132, 135–40
competitive exclusion principle, 225, 227, 245, 247
complex numbers, 128, 140–2
complex of species
  formation of, 248
  of natural enemies, 220, 255, 258, 260–1, 266
  of predators (parasitoids), 219–21, 248, 262, 265
  of prey (hosts), 219, 248, 255, 260
complex-vs-complex interactions, 219–20
conditional reproduction curves, *see* reproduction curves
conditional reproductive rates, *see* reproductive rates
continuous-time process, 35–6, 39, 45, 47, 52, 63, 82–3, 113, 156, 181–2
continuum
  competition–sociality, 131, 139
  scramble–contest competition, 126, 132
correlation
  between climate and population, 170–3
  between reproductive rate and population density, 173
  spurious (nonsense), 176
correlogram, 201–4
covariances, 147, 159, 161, 175
crypsis (mimesis), 26–9

density dependence, 152–4, 167–8, 177, 188
density-dependent
  factors, 145–6, 152
  processes, 153, 163, 168, 173
  regulation, see regulation of populations
density-independent
  factors, 145–6
  processes, 154–5, 158, 173, 175–6
  regulation, see regulation of populations
derivatives, 36, 44–5, 52, 55–6, 64–6, 77, 92–6, 123, 142–3, 217
descriptive models (device), 39, 52, 90–1, 99, 106, 114, 127, 219, 268
detrending (differencing) of series, 204
developmental stadium (stage), 35, 109, 124, 220, 246, 254, 262, 265, 267
difference (recurrence) equation, 182, 218
differential, 36, 46, 55, 83
differential equation, 32–4, 37, 47, 49, 52–4, 88, 182, 217–18
  solution of, 53
discrete experiment, 132–6, 217
discrete-time processes (models, schemes), 35, 39–41, 45, 47, 63, 82–3, 98, 112, 181–2
dispersal, see spruce budworm
distribution, see probability (spatial) distribution
dynamical (time-dependent) processes (pattern), 52, 75, 78–83, 98, 154, 168, 182, 186, 189–95, 205, 217, 232, 237–8, 255, 258, 268

egg/moth (E/M) ratio, 263–6
elimination of competing species, 227–8, 230, 233, 238, 241, 244–7, 269
emigration and immigration, 34, 263–4, 266
empirical ecology, 270
endemic–epidemic dichotomy, see spruce budworm
endogenous processes, 85–9, 181, 192, 197–201, 236, 241–4, 260
  attributes of, 149, 152, 154, 186, 197, 201, 204, 206, 209
  models (curve, form) of, 85–9, 152–5, 172, 177, 181, 202, 206, 212, 215, 218
  parameters of, 85–6, 153, 163, 206, 216
endogenous and exogenous processes, 85–6, 89
ensemble
  averages, 165
  of series, 149, 158, 161–2, 164

environmental influences, 51, 85, 170, 198
environmental resistance, 38
equilibrium density, 49, 62, 84, 152, 154, 167–8, 174, 232–3, 238, 259
  of logistic models, see logistic models
ergodicity, 165
errors, see exogenous influences, random walk
Euler's number and formula, 56, 65, 141–2
exogenous
  processes (factors), 75, 85, 89, 162
  influences, see environmental influences
expectations (expected values), 44, 101, 116–17, 147–54, 159, 164–5, 177–8
exponential rate of (population) increase, see Malthusian
extinction of populations, 48, 50–1, 84, 126, 130, 145, 153, 155, 160, 162, 168, 196, 218, 233

factorial (as mathematical operation), 43, 102
facultative and catastrophic factors, 145
feasibility of population attributes
  ecological feasibility, 83–4, 113, 129, 195, 260
  physical (mathematical) feasibility, 129, 195–6, 233
fecundity, see life cycle in spruce budworm
first principles, 89, 219, 245
fitness, 25, 84, 129, 249
food webs, 177, 219
formulation of models, 41, 44, 108, 110, 183, 186, 225
functional vs regression relationships, 174, 176
fur-return statistics, see lynx cycle

game of trapshooting, 100, 105–10, 113, 128
Gause's law, see competitive exclusion principle
genetic traits, 30, 84, 126, 130
geometric models of competition, 40
  individual-centred, 106, 108, 110
  quadrat-scheme, 106–8
geometric progression, see Malthusian
great tit (*Parus major, minor*), 7–8, 11, 27
Green River Project, see spruce budworm

Hassell model (of single-species population process), 99, 113, 127–8, 137
host (alternative), 248

host complex, *see* complex of species
host–parasitoid interactions, *see* predator–prey interactions
hyperparasitism, 262

immigration, *see* emigration and immigration
increase (in populations)
   innate capacity for, 38
   instantaneous rate of, 36, 48–9
   intrinsic (potential) rate of (natural increase), 38, 48
independent, identically distributed (iid) random numbers, *see* random numbers
index (coefficient) of dispersion (aggregation), 107
industrial melanism, 26–7
infestation map, 255, 258
inflection point of curve, *see* tracing of curves
initial state, 54, 81, 156, 192, 218, 236
initial-state (value) dependence, 218, 236
instantaneous rate of change (increase), *see* rate of change (in population)
intensity of competition, *see* competition
interactions among competing species, 221, 225, 227

lags, *see* autocorrelation, autocovariance, phase lag
l'Hôpital's rule, 57, 65, 96, 118
Life cycle, 40, 253–4, 263
limit cycles, *see* population cycles
logarithmic transformation, 68, 71, 95, 98, 255–8
logarithms
   common, 72–3, 183
   natural, 37, 43, 46, 53, 56, 65, 68, 121, 183, 230, 251
   of negative number, 128, 130, 140
logistic curve, 39, 73, 90
logistic law, 31, 39, 46, 49, 51, 90, 267
logistic models
   classical, 32–3, 46, 63, 68, 82–3, 88, 98, 113, 182
   common version of, 33, 36–7, 47, 49–56, 82
   equilibrium of, 49–50, 52
   generalization of, 98
   Pearl–Reed version of, 33, 36–8, 47, 73–4, 88, 90–1
   Verhulst version of, 33, 37, 45, 52
logistic processes
   continuous-time, 63, 113

discrete-time, *see* discrete-time processes
   generalization of, 69
   theory of, 50, 52, 83, 92, 98, 182
Lotka–Volterra model, 182, 188, 217–18, 241
lynx cycle, 187–8, 190, 221, 266

Malthusian
   geometric progression, 32, 35–8, 50–1, 72, 91, 130–1, 140, 155, 217
   theory of population, 32
matrix
   variance–covariance, 159, 166–7, 179
   the Leslie, 60
measure of aggregation, *see* index of dispersion
micro-niches, 9, 12, 246, 249
mimesis, *see* crypsis
models of competition, *see* geometric models of competition
model parameters
   ecological meanings of, 36, 49, 51, 63, 99, 113, 127, 184–5, 204, 226, 239
   mathematical roles of, 63, 66
moments (statistical), 105, 147, 149–50, 164–5, 204
Moran effect (of population synchrony), 266
Moran plot, *see* reproduction curves
multiple-species interactions, *see* complex-vs-complex interactions

negative binomial distribution, *see* probability distributions
net rate of change in population, 39–40, 45, 61, 121
Nicholson–Bailey model, 218–19
niche segregation, 246–8
niche selection, 25
non-negative entity (counts, quantity, series), 66, 68, 71, 73, 89, 102, 107, 187, 230, 267
nonsense correlation, *see* correlation (spurious)
non-stationary process, *see* stationary process
numerical response, 220

open system, 249
outbreak process, *see* spruce budworm

parasitoid–host interaction, *see* predator–prey interaction

276 · **Index**

Pearl–Reed (logistic) model, *see* logistic models
persistence of populations, 146, 148–54, 158, 162, 164, 167–8, 170, 173
persistent state of populations
    requirements for (stipulations of), 151, 160–2, 166–8, 170–1
    fragile state, 170, 176, 218, 241
phase lag, 220
phase-space representation (orbit), 189–92, 201, 204, 213, 218
Poisson distribution, *see* probability distributions
population (ensemble) averages, *see* averages
population cycles, 192, 260, 262, 265
    broad (or narrow) sense, 193
    chaotic cycles, 82
    limit cycles, 80–4, 189–90, 195, 199, 202, 218, 221
population extinction, 48–51, 84, 126, 130, 145, 153, 155, 160, 162, 168, 196, 218, 233
population oscillation, 76, 80, 153, 189–90, 192–5
population processes, *see* continuous-time processes, discrete-time processes
population synchrony, *see* Moran effect
potential (maximum) reproductive rate, 42, 45, 48–51, 61, 63, 83–5, 88, 98, 109, 121, 134, 148, 168, 183–4, 188, 190, 193, 195–6, 226
potential rate of (population) increase, *see* increase (in populations)
predation, 181, 192
    absence of, 193, 195
    by birds, 26–8, 30, 252, 262
    effect of, 185, 193
    model of, 266
    rate of, 226
    pressure, 29
    visual, 26, 28–30
predator complex, *see* complex of species
predator–prey interaction process, 84, 181–3, 186–7, 196–7, 206, 210, 216–18, 225–6
prey complex, *see* complex of species
probability distributions
    binomial, 43, 99–103, 115
    negative binomial, 99–100, 103–7, 112, 114–15
    normal, 103, 149, 156, 163, 198
    Poisson, 43, 100, 105, 107, 112, 222
    uniform, 43, 149
probability distribution functions
    density function, 103
    mass function, 103–6, 116
profitability, 11, 13, 18–25, 249
profitability curve, 21–3, 184

quadrat scheme (model, system), 106–10, 114, 127

random numbers
    correlated series of, 198–9, 201, 223, 241, 251
    generator, 148, 156, 198
    iid, 148–9, 153–6, 161, 163, 168–9, 175, 198, 200–2, 223, 241, 251
    independent (uncorrelated), 43, 149, 158, 162, 170, 175, 178, 198, 224
    sum of, *see* random walk
random processes, *see* stochastic processes
random search (postulation of), *see* search image theory
random walk, 156–8, 161–2, 168, 170, 175–6, 218
rate of change (in population), 34–7, 50, 54, 67, 74, 130, 147–8, 151–2, 155, 175, 182, 204, 206, 222, 254, 260
    gross rate, 34
    instantaneous rate, 36, 48–9, 217
    net (intergeneration) rate, 35, 39–42, 45, 121
    per-capita rate, 34–5, 37–8, 67–8, 71, 151
    per-unit time rate (of increase), 34, 66–8, 91
regulation of populations, 145–6, 160–1, 168, 171, 175
    biotic(-factor) control theory (school), 145, 168, 172–3
    climatic-control theory (school), 145–6, 169–72
    density-dependent regulation, 162, 167–8, 173–6
    density-independent regulation, 169–73, 175–6
    essence of regulation, 160–1
    testing by regression, 174, 177
reproduction curves, 58–63, 66–88, 123–4, 127, 130–77, 206, 208
    Moran plot, 59–61, 69, 137
    Ricker's curves, 60
    conditional, *see* reproduction surfaces

reproduction surfaces, 206–16, 260
reproductive rate, 40–1
    conditional, 207–8
    formulation of, 185
    potential, *see* potential (maximum) reproductive rate
resource
    depletion (recovery) of, 181, 192
    limit, 51
    requirements, 41, 49, 61, 63
    supply, 40–1, 47, 85, 91

search image (theory), 8–9
series experiment, *see* Utida experiments
sigmoid growth curve, 32, 37, 63, 66–8, 90–1
sigmoidality, 66–7, 72, 90, 267
single-species populations (processes), 60, 84, 98–9, 113, 120, 139, 171, 181, 183, 190, 193, 195–6, 206, 215, 218, 226, 230
slope of curve, *see* derivatives, tracing of curves
sociality (socialness), 128–32, 136–40
spatial distribution, 42, 99, 105–9, 125
species complex, *see* complex of species
spruce budworm, 165
    endemic–epidemic dichotomy, 253–9, 267
    epicentre hypothesis, 253–5, 258
    Green River Project, 252, 254, 260, 263, 265, 267

life cycle, 253–4, 263
moth dispersal, 262–6
outbreaks, 31, 174, 196, 254–62, 266
stationary processes (series, etc.), 164–6, 171, 174, 180, 202, 204
stochastic processes, 31, 146–8, 165, 202, 204, 270
struggle for existence, 225, 245, 249

Taylor–Maclaurin series, 44–5
theoretical ecology, 269–70
time-dependent processes, *see* stochastic processes
tracing of curves, 63–4, 89, 215
    apex (trough), 63–6, 94
    inflection point, 66, 68, 91, 94–5
    slope, *see* derivatives

uniform distribution, *see* probability distributions
Utida experiments, 98–9, 132–4, 217

variance–covariance matrix, *see* matrix
variance-mean ratio, *see* index of dispersion
Verhulst logistic model, *see* logistic models

wandering path, *see* random walk
wildlife
    management, 89, 219
    ten-year cycle, 220